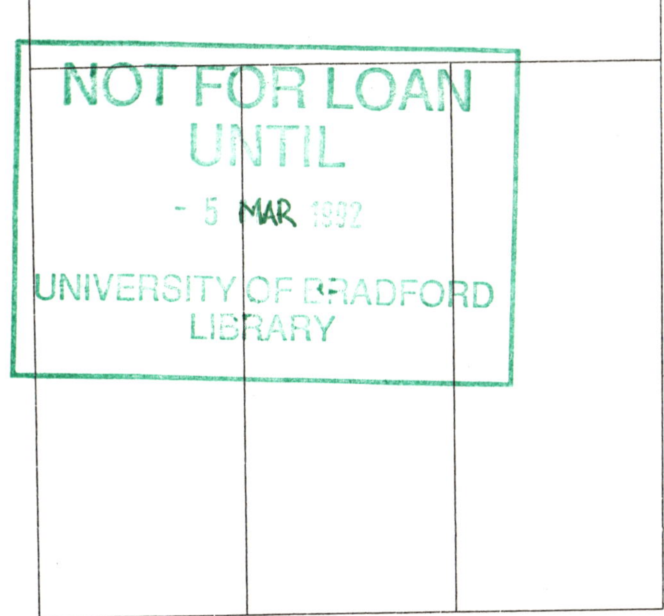
NOT FOR LOAN
UNTIL
- 5 MAR 1992
UNIVERSITY OF BRADFORD
LIBRARY

Prefabricated Vertical Drains: Design and Performance

CIRIA, the Construction Industry Research and Information Association, is an independent non-profit-distributing body which initiates and manages research and information projects on behalf of its members. CIRIA projects relate to all aspects of design, construction, management, and performance of buildings and civil engineering works. Details of other CIRIA publications, and membership subscription rates, are available from CIRIA at the address below.

This CIRIA Ground Engineering Report was written by Professor M B Jamiolkowski and his co-authors under contract to CIRIA. The project arose from studies commissioned by Ente Nazionale per l'Energia, Centro Richerca Idraulica e Strutturale (ENEL–CRIS) who permitted the authors to use the results in this CIRIA project. CIRIA gratefully acknowledges the help of ENEL–CRIS, financial support from the Department, of the Environment for the CIRIA project, and the advice of the following individuals during the project:

Professor T H Hanna	University of Sheffield
Mr F H Hughes	Cementation Piling and Foundations Ltd
Professor A McGown	University of Strathclyde

The authors have asked CIRIA to point out that this text was essentially finalised by 1989; publication was delayed by editorial difficulties beyond their control. Inevitably there have been gains in knowledge since then which are not discussed in this volume.

CIRIA's research manager who instituted this project was Mr J M Head, now of Sir Alexander Gibb and Partners.

CIRIA's research manager for ground engineering is F M Jardine.

CIRIA
6 Storey's Gate
London SW1P 3AU
Tel. 071 222 8891
Fax. 071 222 1708

CIRIA Ground Engineering Report:
Ground Improvement

Prefabricated Vertical Drains: Design and Performance

R D Holtz BS, MSCE, PhD
Professor of Civil Engineering
University of Washington

M B Jamiolkowski, DEng
Professor of Geotechnical Engineering
Politecnico di Torino

R Lancellotta, DEng
Associate Professor of Geotechnical Engineering
Politecnico di Torino

R Pedroni, DEng
Research Engineer ENEL–CRIS

Butterworth-Heinemann Ltd
Linacre House, Jordan Hill, Oxford OX2 8DP

OXFORD LONDON BOSTON
MUNICH NEW DELHI SINGAPORE SYDNEY
TOKYO TORONTO WELLINGTON

First published 1991

British Library Cataloguing in Publication Data
Holtz, Robert D.
Prefabricated Vertical Drains: Design and Performance.
1. Structures. Vertical drainage systems
I. Title II. Jamiolkowski, M. B. III. Pedroni, R. IV. Construction Industry Research and Information Association
V. Series
628.21

ISBN 0 7506 1016 6

Library of Congress Cataloguing in Publication Data
Holtz, R. D. (Robert D.)
Prefabricated Vertical Drains: Design and Performance / R.D. Holtz, M. B. Jamiolkowski, R. Pedroni.
p. cm.— (CIRIA ground engineering report)
— Includes bibliographical references
ISBN 0 7506 1016 6
1. Vertical drains. I. Jamiolkowski, M. B. II. Pedroni, R.
III. Title. IV. Series.
TC974.2.H64 1990
627'.5 — dc20

Typeset by Latimer Trend & Company Ltd, Plymouth
Printed and Bound by Courier International Limited, East Kilbride

Contents

Notation

Symbol	*Units*	*Meaning*
A_f	MN/m^2	Skempton's pore-pressure coefficient at failure
A_w	m^2	cross-sectional area of drain
AOS	μm	apparent opening size of a geotextile
a	m	width of band-shaped drain
a_f		Henkel's pore-pressure coefficient at failure
B_q		pore-pressure ratio
b	m	thickness of band-shaped drain
CR		virgin compression ratio
C_α		secondary compression index
$C_{\alpha\varepsilon}$		secondary compression ratio
C_c		compression index
C_k		permeability change index
C_r		slope of recompression curve ($e - \log_p$) or recompression index
c	m^2/s	coefficient of consolidation
c_h	m^2/s	coefficient of consolidation for horizontal flow
c_u	kN/m^2	undrained shear strength
c_v	m^2/s	coefficient of consolidation for vertical flow
D_{15}	mm	size of the finest 15% of soil particles
D_{85}	mm	size of the finest 85% of soil particles
d_e	m	diameter of equivalent soil cylinder
d_m	m	diameter of mandrel used for drain installation
dQ_1	m^3/s	flow into element of vertical drain
dQ_2	m^3/s	discharge from vertical drain
d_s	m	diameter of smeared zone
d_w	m	diameter of vertical drain
E_{oed}	MN/m^2	constrained (or oedometric) modulus
E_u	MN/m^2	undrained Young's modulus
e		void ratio
e_o		initial void ratio
GR		gradient ratio: result from a soil/geotextile permeameter test (see Chapter 6, Section 6.2 and Figure 38)

G_u	MN/m^2	undrained shear modulus
$H_{1 \ldots n}$	m	measured head heights (see GR (gradient ratio) Chapter 6, Section 6.2 and Figure 38)
H_d	m	drainage height
I_R		rigidity index
i		hydraulic gradient
i_o		initial hydraulic gradient
k	m/s	coefficient of permeability
k_o	m/s	initial coefficient of permeability
k_h	m/s	coefficient of permeability for horizontal flow
k_r	m/s	coefficient of permeability of remoulded soil
k_v	m/s	coefficient of permeability for vertical flow
k_w	m/s	coefficient of longitudinal permeability of the drain
L		well resistance parameter (after Yoshikuni and Nakanodo, 1974)
$L_{1 \ldots n}$	m	lengths over which heads are measured (see GR (gradient ratio) Chapter 6, Section 6.2 and Figure 38)
ℓ	m	characteristic length of drain
m_v	M^2/MN	coefficient of volume compressibility
n		drain spacing ratio ($n = d_e/d_w = r_e/r_w$)
O_{15}	μm	size of opening in a geotextile for which 15% of the openings are smaller
O_{95}	μm	size of opening in a geotextile for which 95% of the openings are smaller
OCR		overconsolidation ratio
P_L	kN	longitudinal tensile force
P_T	kN	lateral tensile force
p_L	kN/m	longitudinal tensile strength per unit width
p_T	kN/m	transverse tensile strength per unit length
q_c	MN/m^2	cone resistance
q_t	MN/m^2	total cone resistance
q_w	m^3/yr	discharge capacity of drain
R		well resistance parameter (after Aboshi and Yoshikuni, 1967)
R_c	m	equivalent radius of cavity
R_r	m	radius of pushing rod
RR		recompression ratio
r	m	radial distance from centreline of drain
r_e	m	radius of equivalent soil cylinder ($d_e/2$)
r_w	m	radius of vertical drain ($d_w/2$)
s		smear zone ratio (d_s/d_w)
$s_{c(t)}$	m	measured consolidation settlement at time t
s_{cf}	m	final consolidation settlement
$s_{(t)}$	m	settlement at time t ($t = 1 \ldots n$)
s_i	m	immediate settlement
s_o	m	intercept on time/settlement curve

T_{80}		time factor to, for example, 80% consolidation
T_h		time factor for consolidation by horizontal drainage
T_{hc}		time factor for consolidation at end of loading ramp
T_v		time factor for consolidation by vertical drainage
t	s	time
t_p	s	time to end of primary consolidation
U		degree of consolidation
U_{hz}		degree of consolidation by horizontal drainage at depth, z
U_p		degree of consolidation in terms of pore-pressure dissipation
U_s		degree of consolidation in terms of settlement
$\bar{U}$		average degree of consolidation
$\bar{U}_h$		average degree of consolidation in terms of horizontal flow
$\bar{U}_p$		average degree of consolidation in terms of pore-pressure dissipation
$\bar{U}_r$		average degree of consolidation for radial flow
$\bar{U}_s$		average degree of consolidation in terms of settlement
u	kN/m²	excess pore pressure (also Δu)
u_{max}	kN/m²	penetration pore pressure
u_o	kN/m²	initial excess, of hydrostatic, pore pressure
u_w	kN/m²	pore pressure
u_{wo}	kN/m²	initial pore pressure
u_T	kN/m²	penetration pore pressure (penetrometer testing)
$\boldsymbol{u}$		normalised excess pore pressure ($=\Delta u_{(t)}/\Delta u_o$)
W_R		well resistance parameter (after Hansbo, 1981)
Z		normalised depth, i.e. depth as a proportion of total depth or thickness of a soil layer
z	m	depth
β		slope of Asaoka's (1978) construction
γ	kN/m³	unit weight
γ_w	kN/m³	unit weight of water
Δu	kN/m²	excess pore pressure
Δu_o	kN/m²	initial excess pore pressure
$\Delta u_{(t)}$	kN/m²	excess pore pressure at a given time
Δu_{max}	kN/m²	excess pore pressure at end of loading ramp
Δu_w	kN/m²	change in pore pressure
$\Delta\sigma_{oct}$	kN/m²	change in octahedral stress
$\Delta\sigma_v$	kN/m²	change in vertical stress
$\Delta\tau_{oct}$	kN/m²	change in octahedral shear stress
ε_L		longitudinal strain
ε_T		transverse strain
ε_v		vertical strain
η		Bjerrum's correction factor
ν		Poisson's ratio (undrained)
ν'		Poisson's ratio (drained)

ρ		normalised radius (r/r_w)
σ	kN/m^2	total stress
σ'	kN/m^2	effective stress
σ_h	kN/m^2	total lateral stress
σ_{vo}	kN/m^2	total overburden stress
σ'_f	kN/m^2	final effective stress
σ'_p	kN/m^2	preconsolidation stress
σ'_v	kN/m^2	effective vertical stress
σ'_{vc}	kN/m^2	effective vertical consolidation stress
σ'_{vf}	kN/m^2	final effective vertical consolidation stress
σ'_{vo}	kN/m^2	effective overburden stress

Introduction

Over the past few decades, more and more civil engineering projects are having to be built on sites underlain by thick deposits of soft cohesive soils. In such conditions, some method of soil improvement is generally required to provide adequate soil-bearing capacity and tolerable post-construction total and differential settlements. These goals may be achieved by precompression, i.e. by preloading the site prior to construction.

Precompression and preloading frequently are used in combination with vertical drains – particularly in very thick deposits of soft cohesive soils – because, otherwise, the time required for consolidation may be unacceptably long, or there may be concern about possible foundation instability. In such cases, the installation of some type of internal drainage system may be the appropriate solution. Traditionally, with one notable exception, these drainage systems used types of vertical drains made of sand. The historical exception is the Kjellman paper wick drain (Kjellman, 1948) which is the prototype for all modern band-shaped prefabricated drains.

Precompression techniques, particularly when used together with vertical drains, are similar to other ground-improvement techniques. They have great potential benefit in a number of situations in geotechnical practice. However, because many practising engineers are not familiar with the advantages and limitations of these methods, or with the design criteria, precompression techniques are not used as widely as they might be. In part, this stems from the inherent difficulties in evaluating time schedules and costs related to their field applications.

This CIRIA Report summarises the current state of the art and practice of prefabricated band-shaped drains.

A number of comprehensive papers and state-of-the-art reports have appeared during the past 15 years dealing with preloading, stability of embankments on soft foundations, and various aspects of soil improvement. Notable are those by Johnson (1970a, 1970b, 1974), Bjerrum (1972), Pilot (1977), Akagi (1979), Schlosser and Juran (1979) and Mitchell (1981). Papers and reports specifically on deep drainage include those in the *Symposium in Print on Vertical Drains*, sponsored by the Institution of Civil Engineers in 1981. Five excellent papers were published in *Géotechnique* (**31**, 1, 1981) and later in *Vertical Drains* (Thomas Telford Ltd). Also important are the *Proceedings* of the Eighth European Conference on Soil Mechanics and Foundation Engineering (Helsinki), with its theme *Improvement of Ground*. Special lectures to that conference, particularly those by Hartikainen (1983), Hansbo (1983c) and van Weele (1983) are recommended highly because they contain good, practical information on the involvement of ground-improvement techniques in foundation engineering. The papers to Helsinki Specialty Session 6, on 'Speeding-up of Consolidation' and the 'General Report' for that session by Jamiolkowski, Lancellotta and Wolski (1983a) are particularly relevant to this Report. Other useful references include that by Magnan (1983) and the comprehensive study by Rixner, Kraemer and Smith (1986).

1 How to use this Report

This chapter outlines the arrangement of the rest of the Report, based on a logically ordered consideration of the use of a vertical-drain scheme for a ground-improvement project. It starts with the assessment of the suitability of such a scheme for the project, then considers ground investigation, selection of drain type, design, installation and the evaluation of results. Under each heading, where appropriate, the reader is referred to those chapters where more detailed information is given.

1.1 How to approach a possible site-improvement and vertical-drain project

In this section, an indication is given to how to decide whether preloading and a vertical-drain system to speed up consolidation is a feasible solution in comparison with other soil-improvement schemes or different foundation types. Figure 1 illustrates the steps taken in foundation design. Soil improvement should be considered relatively early in the design process. Good reasons for using soil improvement rather than going to the next step (e.g. deep foundations) are as follows (Massarsch and Broms, 1983).

(1) Reducing the settlement.
(2) Increasing the bearing capacity.
(3) Decreasing the permeability.
(4) Changing the dynamic response.
(5) Reducing the risk of liquefaction.

In order to help assess which method of ground improvement may be most appropriate, consider the following factors (Mitchell, 1981).

(1) The purpose to which the treated ground will be put. This establishes the level of improvement required in terms of properties such as strength, stiffness, compressibility and permeability.
(2) The area, depth and total volume of soil to be treated.
(3) Soil type and its initial properties.
(4) Availability of materials, e.g. sand, gravel, water and admixtures.

(5) Availability of equipment and skills.
(6) Environmental factors, e.g. waste disposal, erosion, water pollution and effects on adjacent structures and facilities.
(7) Local experience and preferences.
(8) Time available.
(9) Cost.

In general, preloading and vertical drainage are most suitable for predominantly fine-grained, inorganic high water-content, low-strength soils. Typically these sites essentially are normally consolidated or only lightly overconsolidated. Vertical drainage is usually less effective in peats or organic soils, although Fürstenberg *et al.* (1983) and Baranski *et al.* (1986) have reported the successful use of prefabricated drains in such soils.

The type of structure to be supported on the improved ground also affects the applicability of a particular improvement technique, as shown in Table 1. Also important is the question of whether the improvement can be carried out in advance of actual foundation construction.

Preloading with vertical drains requires time for the dissipation of excess pore pressure and the settlements to occur. This may be an important consideration in the overall design. Similarly, because ground improvement can influence the foundation and superstructure (as Hartikainen (1983) and Hansbo (1983c) noted) care should be exercised in comparing costs of various ground-improvement methods with, for example, the alternative of deep foundations, i.e. the comparison should take into account total project costs, not just the prime cost of the ground improvement or piling.

Finally, as Hartikainen (1983) noted, the risks involved in certain soil conditions have to be kept in mind when choosing the soil-improvement method.

When the soil conditions are worse than expected (as is often the case) the vertical drains may not perform as intended. In such a situation, there are several options as follows.

(1) Installing additional drains (decreasing the drain spacing).
(2) Increasing the level of surcharge.
(3) Increasing the time for preconsolidation.
(4) Reducing the acceptance criterion (which, in effect, means a modified foundation and/or structure).

The last choice is clearly the worst one. However, as the possibility always exists, prudence suggests that the limits of applicability of the chosen method should not be approached too closely.

1.2 Site investigation

The types of soils and site conditions in which soil improvement with preloading and vertical consolidation are likely to be most effective were mentioned above. The

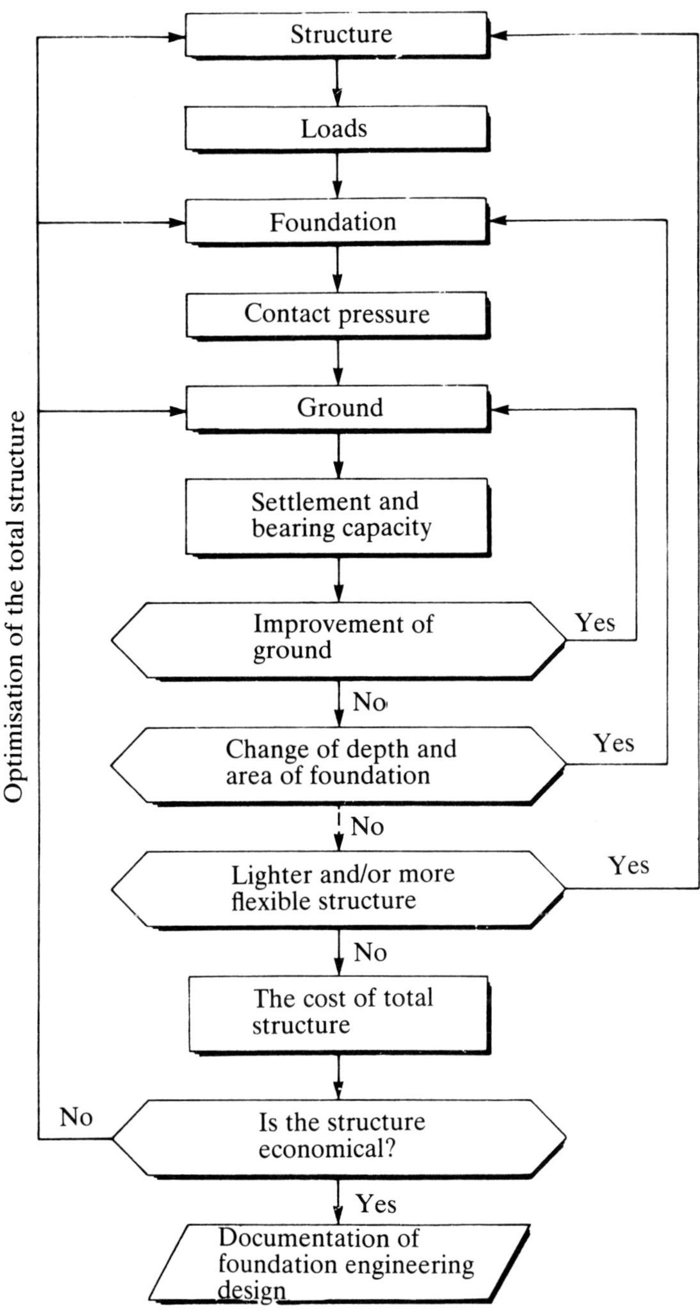

Figure 1 Flow chart illustrating the foundation design process (Hartikainen, 1983)

Table 1 Applicability of foundation soil improvement for different structures on shallow foundations on soft alluvial deposits and old inorganic fills (adapted from West (1975) by Mitchell (1981))

Category of structure	*Structure*	*Permissible settlement*	*Load intensity (usual bearing pressure required) (kN/m²)*	*Probability of advantageous use of soil improvement techniques*	
				Soft alluvial deposits	*Old, inorganic fills*
Office/ apartment	High rise: more than 6 storeys	Small <25–50 mm	High (300)	Unlikely	Low
(frame or load-bearing construction)	Medium rise: 3–6 storeys	Small <25–50 mm	Moderate (200)	Low	Good
	Low rise: 1–3 storeys	Small <25–50 mm	Low (100–200)	Good	High
Industrial	Large span with heavy machines, cranes; process plants; power plants	Small <25–50 mm; differential settlement critical	Variable (high local concentrations to >400)	Unlikely	Low
	Framed warehouses and factories	Moderate	Low (100–200)	Good	High
	Covered storage, store-rack systems, production areas	Low to moderate	Low (<200)	Good	High
Others	Water and waste water treatment plants	Moderate; differential settlement important	Low (<150)	High	High
	Storage tanks	Moderate to high, but differential may be critical	High (up to 300)	High	High
	Open storage areas	High	High (up to 300)	High	High
	Embankments and abutments	Moderate to high	High (up to 200)	High	High

preliminary subsurface information sufficient for the appraisal of the viability of soil improvement or other foundation solutions is unlikely to be enough for detailed analysis and design of the vertical-drain system. The additional programme of sampling and *in-situ* testing depends on the size of the project and on the preliminary assessment of whether the clay deposit is likely to possess a strong macrofabric (see Chapter 5, Section 5.1).

1.3 How to select the drainage system

If it has been decided that preloading and vertical drains are likely to be feasible, the sections and tables shown in Table 2 should be consulted in deciding which types of drains should be used.

Table 2 Selection of drain types

Type	*Refer to:*
Sand drains	Section 2.2 and Table 7 Section 2.3 Section 2.5 and Table 9
Prefabricated drains	Section 2.2 and Table 8 Section 2.4 Section 2.5 and Table 9

The engineer should also be sure to consult the major references given in the Introduction.

1.4 How to design the drains

Whether conventional sand drains or prefabricated band-shaped drains are chosen, the analytical and geotechnical aspects of design are similar (see Chapter 2, Section 2.5). The site plan, soil profile, and certain soil properties are needed as input to the design equations (Table 3).

Once this information is available, apply the design equations (see Chapter 3, Section 3.2). In order to save programming and calculation time, note that design charts and nomographs have been prepared by, for example, Hansbo (1979) and Gersevanov and Rathmayer (1984).

(1) Input:
   (a) soil properties;
   (b) boundary and drainage conditions;
   (c) desired degree of consolidation;
   (d) time available for consolidation;
   (e) drain installation pattern assumed (square or triangular) (see Chapter 3, Figure 5);
   (f) equivalent diameter (if a band-shaped drain) (see Chapter 6, Section 6.7).

(2) Output:
   (a) drain spacing.

Note: Spacing may be varied to optimise the design. Closer spacings increase installation costs. (See Hansbo, 1983c, for a discussion of costs of vertical-drain systems. Hartikainen (1983) also gives some data on vertical drains relative to land costs in Helsinki, lime columns, excavation and replacement.)

Table 3 Data needed for design

Topic	*Refer to:*
Essential matters	
Area to be drained	Site plan
Length of drains	Soil profile and applied loads
General information	Chapter 4 and Table 13
Specific geotechnical information	Section 5.1
Consolidation and permeability:	Section 5.2
laboratory	Section 5.3
field	Sections 5.4 and 5.5
back analysis	Sections 5.6, 5.7, 5.8 and 5.9
Stress history	Section 5.10
Macrofabric and drainage boundaries	Section 5.11
Other important matters	
One-dimensional, compared with three-dimensional, effects (qualitative only)	Section 5.12
Stress–strain and shear strength (important for stability of preload fills)	Section 5.13
Matters to be considered at least qualitatively	
Secondary compression	Section 5.14
Installation effects	Section 5.15

If additional analytical refinements are desired, then reference to the appropriate parts of Chapter 3 should be made as shown in Table 4.

Table 4 Practical matters affecting analysis

Topic	*Refer to:*
Smear	Section 3.3
Well resistance	Section 3.4
Combined radial and vertical drainage	Section 3.5
Time-dependent loading	Section 3.6

If prefabricated band-shaped drains are selected, their material characteristics have to be appropriately considered, as outlined in Table 5.

Table 5 Material characteristics of band-shaped drains

Topic	*Refer to:*
General information	Sections 2.2, 2.3, 2.4, 6.1, 6.2, 6.7, 6.8
Transverse permeability (filter sleeve)	Sections 6.3, 6.4
Discharge capacity (lateral stress, drain folding, siltation)	Section 6.5
Mechanical properties	Section 6.6
Durability	Section 6.8
Evaluation and specifications of drains	Section 6.9

1.5 How to install the vertical drains

The details of most foundation construction operations are usually left to the contractor. However, with vertical-drain projects, the designer has to be concerned with some of the construction and installation details. A list of items to be considered by the designer is given in Table 6.

Table 6 Aspects of installation and construction

Topic	*Refer to:*
Drain installation method	Sections 2.2, 2.3, 2.4
Site preparation*	
Working platform*	
Instrumentation to control fill placement*	Hanna (1973, 1985)
– piezometers	
– settlement platforms	
– inclinometers	
– plan and elevation surveys	
Fill placement rates, stages etc*	
Removal of surcharge*	
Alternative surcharge (e.g. vacuum method)*	

*These matters are not specifically dealt with in this report: they will depend upon the particular programme or nature of the project or site (Section 1.6)

1.6 How to evaluate the results

As with virtually all designs in foundation engineering, it is very useful for the designer to know how well the foundation system has performed after construction. Visual inspection by the design engineer or his/her representative of all the aspects of the construction of vertical-drain projects (site preparation, drain installation, fill placement, surcharge removal, etc.) is essential. Geotechnical instrumentation to monitor the excess pore pressures and the vertical (and possibly the horizontal) displacements is essential for most large projects, and for any project, however small, in which fill failure would be disastrous. Good information on various aspects of geotechnical instrumentation is given by Hanna (1973; 1985).

Monitoring should be not only during construction, but also afterwards for as long as is practicable. The objective of the long-term monitoring is to verify settlement predictions made during the design phase.

2 Vertical drains

2.1 History

The primary purpose of vertical drains is to accelerate the consolidation time by shortening the drainage path. An additional advantage also comes from the fact that most natural soft-clay deposits are more-or-less anisotropic with respect to their flow properties. Typically, the coefficient of permeability for horizontal flow, k_h, is higher than that for flow in the vertical direction, k_v. Therefore, when horizontal flow predominates, vertical drains generally give an additional benefit to the rate of primary consolidation in comparison to the effects of a purely geometrical change of the drainage boundaries.

The American engineer, D. J. Moran, proposed the use of sand drains as a means for deep soil stabilisation in 1925, and he received a US patent on the concept in 1926 (Johnson, 1970a). The first practical sand-drain installations were constructed in California a few years later (Porter, 1936).

In Sweden, also in the mid to late 1930s, Kjellman began experiments and obtained patents on the first prototype of a prefabricated drain made entirely of cardboard. Details of the 'Kjellman wick drains' were first published in English only after the Second World War (Kjellman, 1948).

According to O. Wager (who worked with Kjellman during the early days of the development of the wick drain) there were problems with the first prototype prefabricated drains. It was soon discovered that the wicks were subject to undesirably rapid deterioration, particularly near the top of the drained clay layers. Treatment of the paper to retard the rotting was objectionable, because it reduced the permeability of the drain significantly (from about 10^{-5} to 10^{-7} m/s) and the retardant they used was a compound of arsenic and, consequently, highly toxic.

Even with these difficulties, Kjellman wick drains have been used occasionally in both Europe and Japan during the past 40 years. However, until the early 1970s, the vast majority of vertical drains installed in the world were sand drains.

In 1971, Wager improved on the Kjellman wick by using a grooved plastic (polyethylene) core in place of the cardboard one. This drain was called the Geodrain and the first models utilised Kraft paper filters. Later models were provided with non-woven textile filters.

In the last ten years, a new frontier seems to have opened for vertical drains. A large number of prefabricated drains, at present more than fifty different types, have appeared on the market. This competition has decreased the cost of the drains appreciably. Installation procedures, too, have been improved. They now permit rapid, simple installations to depths approaching 60 m at rates up to 1 m/s.

2.2 Types, materials and installation procedures for vertical drains

The various types of vertical drains, their common installation methods and pertinent geometric characteristics are given in Table 7. Specific dimensions and materials for a number of common prefabricated drains are listed in Table 8.

Most of the prefabricated drains in Table 8 are similar to the original Kjellman cardboard wick in that they have a central core, typically with grooves, studs or channels for water transport. Surrounding the core is an outer covering which serves as a filter to prevent soil particles from entering the core and possibly decreasing the discharge capacity of the drain. The core also supports the filter during installation and resists the *in-situ* soil pressures.

Filters are made either of 'paper' or of a non-woven fabric or geotextile, although a few drains have both filter and core made of the same material in a type of unitised construction.

Table 7 Types of vertical drains, common installation methods and typical geometric characteristics (after Jamiolkowski, Lancellotta and Wolski, 1983a)

Drain type	*Common installation methods*	*Drain diameter* (m)	*Typical spacing* (m)	*Maximum length* (m)
Sand drain	Driven or vibratory closed-end mandrel (displacement type)	0.15–0.6	1–5	⩽ 30
Sand drain	Hollow stem continuous-flight auger (low displacement)	0.3–0.5	2–5	⩽ 35
Sand drain	Jetted (non-displacement)	0.2–0.3	2–5	⩽ 30
Prefabricated sand drains ('sandwicks')	Driven or vibratory closed-end mandrel; flight auger; rotary wash boring (displacement or low displacement)	0.06–0.15	1.2–4	⩽ 30
Prefabricated band-shaped drains	Driven or vibratory closed-end mandrel (displacement or low displacement)	0.05–0.1*	1.2–3.5	⩽ 60

*Equivalent diameter.

Table 8 Dimensions and materials of some prefabricated drains

Drain type	Dimensions		Materials	
	Width (mm)	Thickness (mm)	Filter	Core
Kjellman	100	3	Cardboard	Cardboard
Alidrain	100	6.1	Geotextile	Polyethylene
Amerdrain	92	10	Geotextile	Polypropylene
Bando	96	2.9	Paper	PVC
Castleboard	93	3.2	—	Polyolefin
Colbond	300	4	Geotextile	Polyester
Desol	95	2	—	Polyolefin
Geodrain	98	4	Paper	Polyethylene
Geodrain	96	3.5	Geotextile	Polyethylene
Hitek	100	6	Geotextile	Polyethylene
Mebradrain	95	3.2	Paper	Polypropylene or
Mebradrain	95	3.4	Geotextile	polyethylene
OV Drain	103	—	Geotextile	Polyester
Solpac Charbonneau	105	—	Geotextile	Polyester
Tafnel	102	6.9	—	Polypropylene

2.3 Displacement, augered, and jetted sand drains

The comments in this and the following sections are based primarily on studies by Rowe (1968), Casagrande and Poulos (1969) and Johnson (1970a), as well as a few recent investigations.

Driven, displacement sand drains are still in wide use, because of their simplicity and relatively low cost. However, their installation causes very severe disturbance to the surrounding soil, often with the following highly undesirable effects on the overall performance of the soil-drain system.

(1) The shear strength of the foundation clay is reduced, at least temporarily. This reduction is related directly to clay sensitivity.
(2) A highly disturbed 'smear' zone is formed around each drain.

Within this zone, any anisotropy with respect to the coefficient of permeability ($k_h > k_v$) is almost completely eliminated. Thus, the pore-water flow is controlled by the isotropic coefficient of permeability of the remoulded clay, k_r, which is typically lower than the coefficient of permeability for vertical flow, k_v, of the undisturbed natural soil.

Among the field studies of disturbance is the analysis of twelve well-documented case records by Akagi (1977a, 1977b) which showed that the shear-strength decrease caused by the installation of displacement-type drains is generally regained by a rather rapid dissipation of the excess pore pressure generated by the installation. Despite this, the use of displacement sand drains is subject to many limitations. They should be avoided particularly when the following conditions apply.

(1) The clay sensitivity exceeds 4–6.

(2) The soil deposit possesses a highly developed macrofabric, associated with high anisotropy of the permeability coefficient.
(3) Stability is critical for the design.
(4) The load transmitted to the foundation is small and the soil is overconsolidated.

Item (4) above is also relevant to other types of vertical drains, not just displacement sand drains.

Sand drains installed by means of a continuous-flight hollow-stem auger probably cause some soil displacement and disturbance, the magnitude of which is intermediate compared to full-displacement and jetted sand drains.

Jetted sand drains minimise the disturbance and the smearing of the surrounding clay because they replace soil with little displacement. Therefore, they are particularly appropriate for sites where driven, displacement-type drains should be avoided. However, their installation is quite complex, and requires an experienced contractor and very close supervision. This installation method can cause difficulties in places where the drains have to penetrate through stiff clays or coarse granular layers, and where environmental constraints complicate the disposal of the excavated clay and jetted water.

Singh and Hattab (1979) conducted laboratory experiments of radial drainage to central model sand drains of different shapes, spacings and installation methods. They rated the various methods, in order of decreasing efficiency, as follows.

(1) Closed-end cross-shaped mandrel.
(2) Jetted.
(3) Augered.
(4) Closed-end circular mandrel.
(5) Closed-end star-shaped mandrel.
(6) Open-end circular mandrel.

This work, while comprehensive as far as sand drains are concerned, unfortunately did not include the mandrel shapes commonly used for the installation of prefabricated drains.

2.4 Prefabricated drains

As shown in Table 7, the prefabricated drains or sandwicks (Hughes and Chalmers, 1972) and band-shaped drains, have much smaller dimensions than ordinary sand drains. Although other methods have been used, the most common installation procedure is by closed-end mandrel. Thus, the amount of soil disturbance caused by installation is determined by the size and shape of the mandrel, as well as by the dimensions and shape of the detachable shoe or anchor located at the mandrel tip. Figure 2 shows in cross-section three typical mandrels for band-shaped drains. Two typical shoes are shown in Figure 3.

The shoe may be a small piece of sheet metal bent into a V shape (Figure 3a), or sometimes a small piece of drain. While the mandrel is still up about 300–400 mm

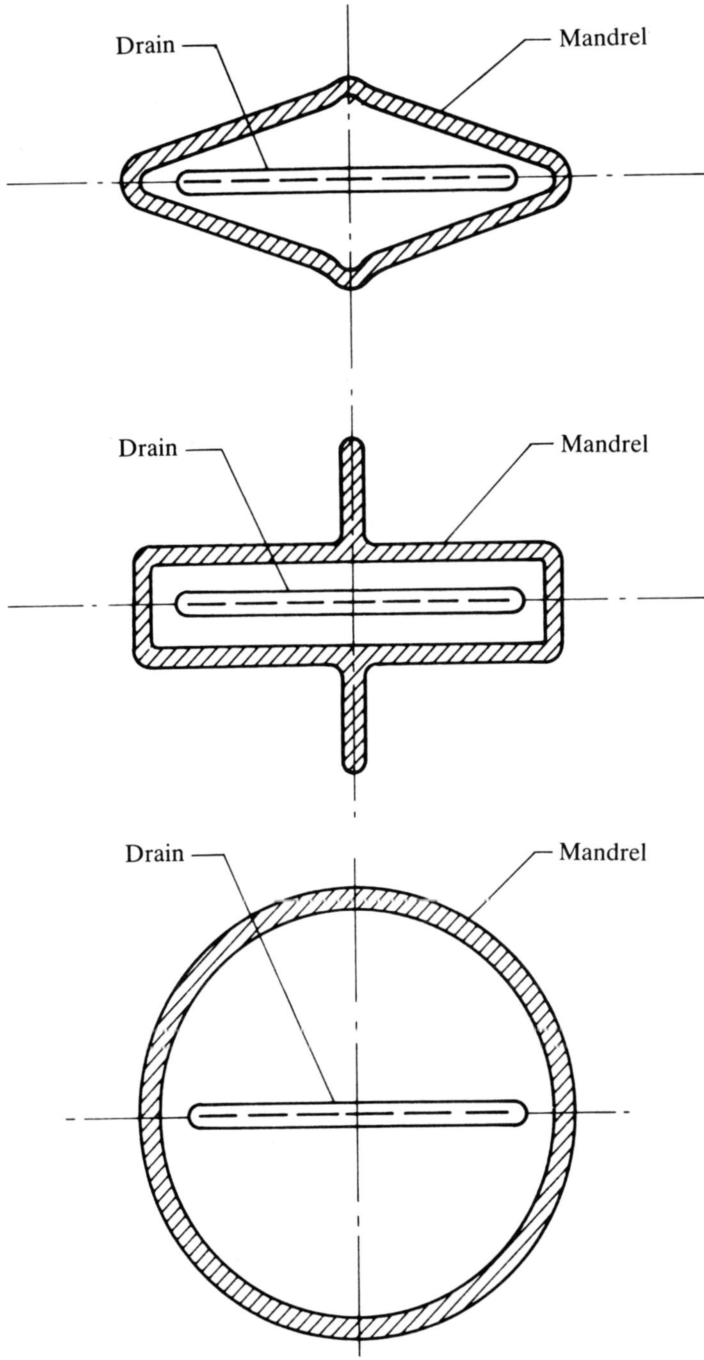

Figure 2 Typical mandrels for band-shaped drains

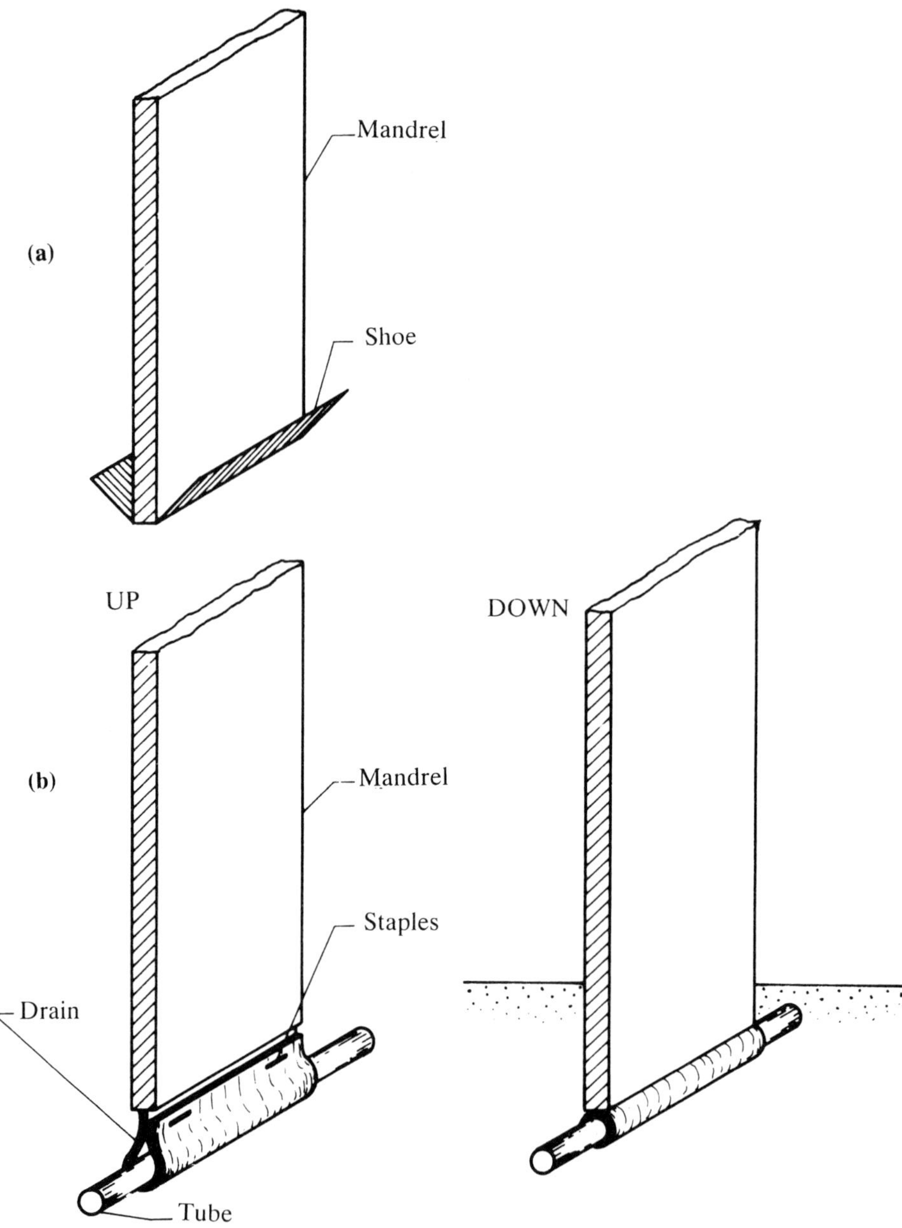

Figure 3 Typical detachable shoes used with prefabricated-drain mandrels

above the ground surface, some drain is pulled out of it, wrapped round a short tube and either stapled or tucked-up into the mandrel (Figure 3b). When the mandrel is lowered to the ground surface at the start of insertion, the drain and tube are pulled snugly up into the mandrel end. The tube may be a short length of metal or plastic pipe, a short wood dowel, or even a rolled-up section of the drain itself. The length of the tube is usually only a few millimetres longer than the width of the mandrel.

The detachable shoe serves the following two purposes.

(1) It acts as an anchor to keep the drain at the desired depth as the mandrel is withdrawn after insertion.
(2) It prevents soil from entering the mandrel and plugging it during drain installation.

Hansbo (1979), McGown and Hughes (1981) and Droy, McGown and Mann (1986) show examples of different disposable end shoes and drain anchors. As far as is known, there have been no definitive studies of the effects of different types of shoes on smear and soil disturbance, although Droy, McGown and Mann (1986) speculate about the disturbance effects of a special end shoe and particular installation procedures used on a drainage site near Glasgow.

The mandrels usually are simply pushed into the soil by a system of rollers, a chain drive, or a gear system. Sometimes vibration is also used (McGown and Hughes, 1981). Whether installation is by static penetration or vibratory driving seems to have no influence on the rate of consolidation, although the pore-pressure responses during the installation appear to be different (Hansbo, Jamiolkowski and Kok, 1981).

The degree of disturbance associated with the installation of prefabricated drains is appreciably less than in the case of conventional displacement sand drains because the volume of the installed drains with respect to the volume of the treated soil is smaller. Therefore, prefabricated drains are appropriate for sensitive clays and soils having high degrees of anisotropy of the permeability coefficient. The advantage of prefabricated band-shaped drains lies mainly in the simplicity and high speed of the installation procedure, which result in relatively low unit installation costs.

Undoubtedly, all these advantages have contributed greatly to the recent increase in the number and usage of prefabricated drains.

2.5 Soil and drain parameters required

The parameters governing the design and performance of precompression, with or without vertical drains, on soft cohesive deposits are as follows.

(1) Consolidation coefficients for horizontal flow, c_h, and vertical flow, c_v.
(2) Stress history of the deposit (preconsolidation pressure, σ'_p, and overconsolidation ratio, OCR).
(3) Extent of the smeared zone, d_s.
(4) Permeability coefficient of the remoulded clay within the smeared zone, k_r.

In conjunction with vertical drains a fifth needs to be considered.

(5) Discharge capacity of the drains, q_w, and its variation with total lateral stress, σ_h, and time, t.

Table 9 The relative importance of design parameters for vertical drains

	Soil characteristics (c_h, σ'_p)	*Smear* (d_s, k_r)	*Drain discharge capacity* (q_w)
Displacement sand drains	Essential	Essential	Not important*
Jetted sand drains	Essential	Not important (?)	Not important†
Prefabricated band-shaped drains	Essential	Important	Essential for long drains

*May become important for sand drains of small equivalent diameter (d_w < 200 mm) whose length exceeds 20–25 m.
†May become important for drains where the sand backfill is contaminated by fines.

The relative importance of these parameters in relation to the type of vertical drains is given in Table 9.

How these parameters are obtained, and the methods of including them in designs for prefabricated drains, are discussed in detail in Chapters 3–6. In addition, factors such as filter permeability, mechanical properties of drains, and drain durability and longevity are also discussed.

3 Consolidation theories for band-shaped drains

The consolidation of a clay stratum is a transient-flow problem in a porous medium. The solution of the problem for a saturated soil requires the six following basic aspects to be considered.

(1) Equilibrium of the porous body.
(2) Strain compatibility.
(3) Interaction between pore water and soil skeleton.
(4) Conservation of mass of the water.
(5) State relationship for the pore water.
(6) Dynamic equilibrium of the pore water.

Relationships (1) and (2) come from continuum mechanics, and they require no additional comments.

The interaction in relationship (3) between the pore water and the soil skeleton expresses how the applied external stresses, the pore-water pressure and the effective stresses are related. The relationship used is the well-known principle of effective stress (Terzaghi, 1923, 1936; Skempton, 1960).

The conservation of mass (relationship (4)), usually expressed as the continuity relationship, requires that the net inflow (or outflow) of water into a given soil elemental volume be equal to the increase (or decrease) of the volume of the same element. The change in soil volume may arise from changes in both the void ratio and the water density. To account for this latter condition, a state equation implied in relationship (5) (i.e. a relationship between the density and the pressure and temperature of the pore water) is required. Finally, a statement of the influence of the applied stresses on the pore-water flow – a dynamic equilibrium relationship (6) analogous to Newton's second law of motion – is also needed because the consolidation problem cannot be solved by the continuity and state equations alone.

Fluids are viscous; therefore, the shear-stress components also have to be included in the equilibrium equations. Their general formulation is known as the Navier–Stokes equations of motion. Their nonlinear nature makes it impossible to solve these equations for general boundary and initial conditions.

This very complex aspect of the consolidation problem has been taken care of by Darcy's (1856) law which, in this case, can be viewed as a relationship statistically equivalent to the dynamic equations of motion. The great advantage of using Darcy's law instead of the general Navier–Stokes equations is that the viscous resistances are implicitly taken into account. Also, the so-called 'surface velocity' is the result of the gravity and pressure forces and the viscous resistances applied to the pore water. In this context, Darcy's law plays an important role in the solution of flow problems in porous media, which is why so much attention has been focused on the verification of its validity.

Some authors (e.g. Hansbo, 1960) suggest the possibility of a nonlinear gradient/velocity relationship. Others (Miller and Low, 1963) postulate the existence of a threshold gradient below which no flow occurs. On the other hand, Olsen (1965) and Mitchell (1976) confirm the validity of a linear relationship between the surface velocity and the hydraulic gradient. Tavenas, Jean, Lablond and Leroueil (1983) have reported the results of extensive research on the permeability of natural soft clays. Tests were performed in triaxial and oedometer equipment which had been modified to eliminate leakage. Figure 4 shows that the flow is continuous under decreasing hydraulic gradients. If a threshold gradient existed, the results would show an abrupt change in the flow curve. From these results, Tavenas, Tremblay and Leroueil (1983) reached the following conclusions.

(1) Darcy's law is valid in natural soft clays under gradients ranging from 0.1 to 50.

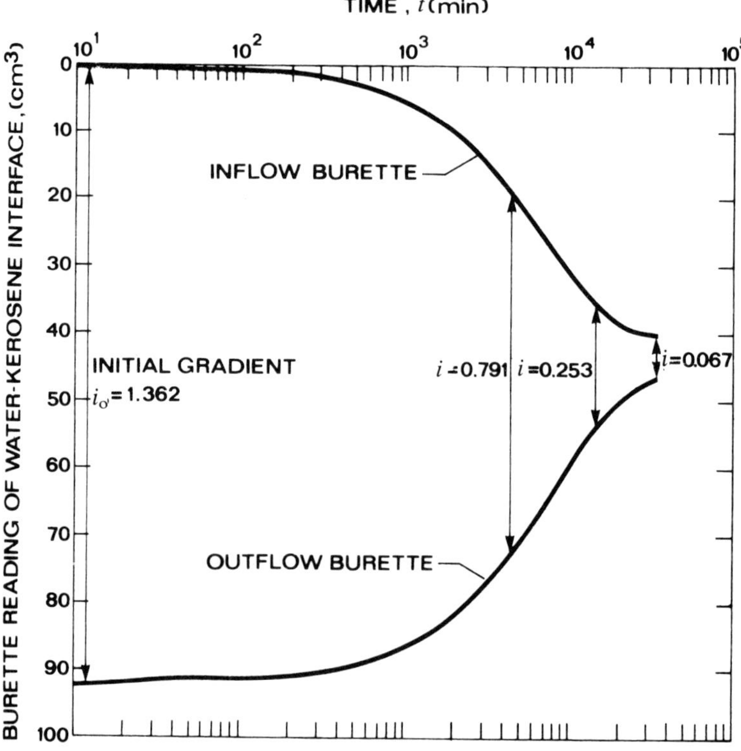

Figure 4 Results of a triaxial permeability test (Tavenas, Tremblay and Leroueil, 1983)

(2) The development of primary and secondary volume changes in clay specimens make it almost impossible to verify the validity of Darcy's law at smaller gradients.

3.1 General theory of consolidation

The basic aspects listed at the beginning of this chapter are taken into account in the complete theory of consolidation first derived by Biot (1941). This exact theory involves four equations (i.e. the water-flow equation through the soil, and equilibrium equations in terms of displacements) with four unknowns, i.e. the pore-water pressure and the displacement components along the coordinate axes.

The complexity of these equations means that only a few solutions are available. Some of these are listed in Table 10. In the context of this Report, the two aspects below have to be considered.

(1) Consolidation in the presence of vertical drains is a problem which, strictly speaking, requires the use of the Biot theory, whereas the theory commonly employed is the Terzaghi (1925) and Rendulic (1937) simplified analysis of the diffusion of water through a porous medium.
(2) The quantity of practical engineering interest – the degree of consolidation in terms of settlement, U_s – may be obtained only by the Biot theory, because the Terzaghi–Rendulic diffusion theory gives the average degree of pore-pressure dissipation, $\bar{U}_p$, which is, in general, different from $\bar{U}_s$.

However, with regard to point (2), it should be stressed that a comparison between the solutions of U_s in Table 10 and the values of U_p obtained from the simplest diffusion theory shows that U_p is never outside the two extremes of U_s for values of drained Poisson's ratio ν' between 0 and 0.5 given by the Biot theory. Thus, the calculated value of U_p is a reasonable practical approximation of U_s with its accuracy

Table 10 Available solutions of the Biot theory for problems of practical engineering interest

Type of problem	*References*
Uniformly loaded strip on semi-infinite soil	Biot (1941); McNamee and Gibson (1960); Schiffman, Chen and Jordan (1969)
Uniformly loaded rectangle on semi-infinite soil	Gibson and McNamee (1957)
Uniformly loaded circle on semi-infinite soil	De Jong (1957); McNamee and Gibson (1960); Schiffman and Fungaroli (1965)
Uniformly loaded circle on a finite layer	Gibson, England and Hussey (1967)
Circular loading on a semi-infinite soil overlain by a permeable layer	Mandel (1957, 1961)

increasing with decreasing layer depth (Davis, 1969; Viggiani, 1970, and Davis and Poulos, 1972).

Moreover, both the Biot and the Terzaghi theories involve significant simplifications of the soil properties. The stress/strain relationship is assumed to be linear and the coefficient of consolidation to be constant over the range of effective stresses encountered in the field. These oversimplifications may lead to significant errors, as shown by Mesri and Rokhsar (1974) and Olson and Ladd (1979). However, classical Terzaghi theory is still used extensively to solve consolidation problems with or without vertical drains, not because of the supposition that it works reasonably well, but as Olson and Ladd (1979) pointed out, for the following reasons.

(1) It is simple to use.
(2) The major sources of uncertainty are in the assessment of the soil properties and the macrofabric of the deposit rather than in the oversimplifications implicit in the theory.
(3) There are few well-documented case studies comparing predicted with measured consolidation behaviour.

It is in this context of the available consolidation theories and their limitations that the theories for vertical drains have to be evaluated. It is why so much attention is given to the assessment of soil properties and to the drain characteristics. It is also why the use of numerical programs is suggested to solve problems where the soil behaviour is expected to be highly non-linear over the range of stresses of practical interest and when large variations of the consolidation (or permeability) coefficient or large strains are expected.

3.2 Theory for pure radial drainage

Vertical drains are used for speeding-up the consolidation of a clay layer by adding radial drainage to the vertical. The drain spacing is typically much less than the vertical drainage distance; thus, radial drainage is generally the dominant component in the consolidation process. Radial drainage becomes still more important as the consolidation coefficient for horizontal flow, c_h, is usually greater than the coefficient for vertical drainage, c_v. The triangular drain pattern (Figure 5a) has been shown (Barron, 1948) to be the most economical one, even if a square pattern (Figure 5b) can also be used for practical reasons (Kjellman, 1948). In both cases, the boundary conditions of the problem to be analysed refer to an equivalent soil cylinder of diameter, d_e (Figure 5c), having an impermeable outside vertical surface and an inner cylindrical drain. If the drain is band-shaped (Figure 6), Hansbo (1979) suggested that the equivalent diameter, d_w, should be that of a cylinder having the same circumference, or (Equation (1)):

$$d_w = [2(a + b)]/\pi \tag{1}$$

where a and b = width and thickness, respectively, of the band-shaped drain (see also Chapter 6, Section 6.7)

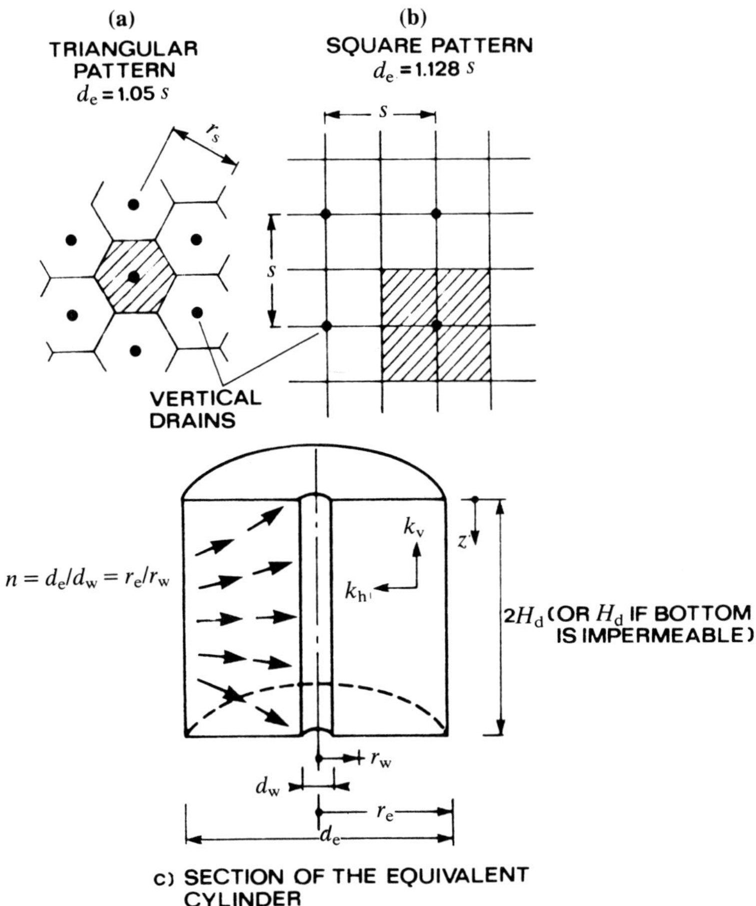

Figure 5 Different drain patterns: equivalent cylinder

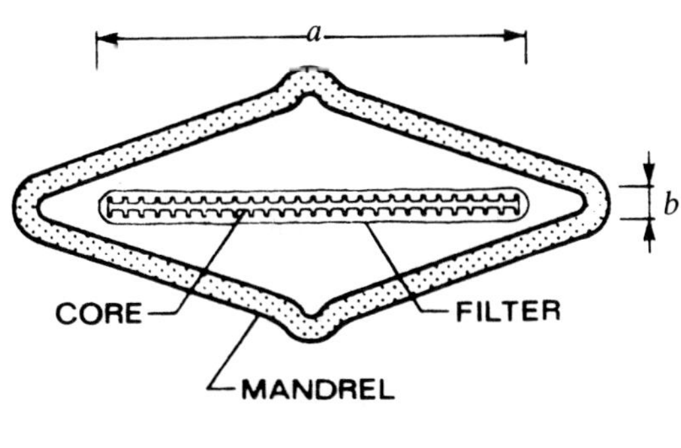

$$d_w = [2(a + b)]/\pi$$

Figure 6 A typical cross-section of a band-shaped drain

The available solutions for vertical drains use all the simplifying assumptions of Terzaghi's uncoupled one-dimensional consolidation theory (Taylor, 1948). They also refer to the following boundary conditions which represent the following two possible field situations.

(1) *Free vertical strain.* The assumption is that the vertical surface stress is constant during the consolidation process and, thus, that the resulting surface displacements are non-uniform.
(2) *Equal vertical strain.* The assumption here is that the vertical surface displacements are constant throughout the drained area and, thus, that the resulting vertical stress at the surface is non-uniform.

Actual field situations are somewhere in-between, depending on the ratio of d_e/d_w, the thickness of the sand blanket, and the ratio of the axial stiffness of the drain to that of the surrounding soil (Holtz and Holm, 1972).

For pure radial drainage, the governing equation in cylindrical coordinates is as shown in Equation (2).

$$\frac{\partial u}{\partial t} = \left(\frac{1}{r} \cdot \frac{\partial u}{\partial r} + \frac{\partial^2 u}{\partial r^2} \right) c_h \qquad (2)$$

where u = excess pore pressure at any point and any time
t = time after an instantaneous increase of the total vertical stress
r = radial distance of the considered point from the centre of the drained soil cylinder
c_h = coefficient of consolidation for horizontal flow, defined as $c_h = (k_h\ E_{oed})/\gamma_w$
k_h = coefficient of permeability for horizontal flow
γ_w = unit weight of water
E_{oed} = constrained or oedometric soil modulus in the relevant stress range

The solution for pure radial drainage and equal strain boundary conditions was first derived by Rendulic (1936). Another theoretical solution for the same case was developed by Kjellman (1937).

Barron (1948) – in the first exhaustive treatment of the problem – gave solutions for both boundary conditions. He also considered smear and well-resistance effects for the case of equal vertical strain. More recently, Yoshikuni and Nakanodo (1974) developed a solution for the free strain boundary with a drain of finite permeability, and Schiffman (1958) and Olson (1977) presented solutions which account for variable loading.

A comparison between Barron's (1948) free and equal strain solutions indicates that both yield almost the same average degree of consolidation for values of the drain spacing ratio, $n = d_e/d_w$, greater than 5 (which is usual) and time factor, $T_h > 0.1$ (Figure 7). This justifies the use of the simpler equal strain solution, which is given in Equation (3).

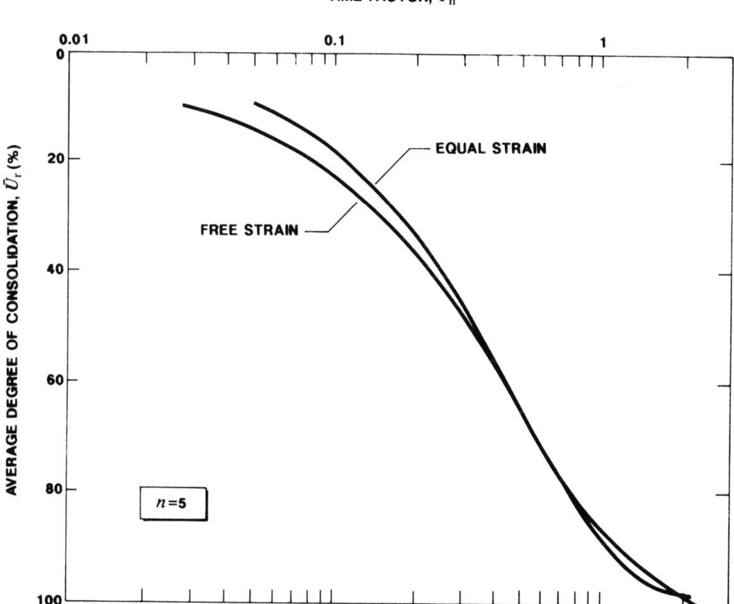

Figure 7 Average degree of consolidation versus time factor for free strain and equal strain boundary conditions; radial inflow tests with the drain spacing ratio = 5 (after Trautwein, 1980)

$$u=\frac{u_o}{r_e^2 \cdot F(n)}\left[r_e^2 \log_e\left(\frac{r}{r_w}\right)-\frac{(r^2-r_w^2)}{2}\right]\exp(\lambda) \tag{3}$$

where u = excess pore pressure
u_o = initial excess pore pressure
r_e = radius of equivalent soil cylinder ($d_e/2$)
r_w = radius of drain ($d_w/2$)

$\lambda = -8T_h/F(n)$

$T_h = (c_h t)/d_e^2$

n = the drain spacing ratio $= (d_e/d_w)$

and

$$F(n)=\frac{n^2}{(n^2-1)} \cdot \log_e(n) - \frac{(3n^2-1)}{4n^2} \tag{4}$$

In this case, the average degree of consolidation with respect to radial flow becomes:

$$\bar{U}_h = 1-\exp\left[\frac{-8T_h}{F(n)}\right] \tag{5}$$

3.3 Smear effect

Equations (3) and (5) assume that the installation of the drain does not change the properties of the surrounding soil. In practice, drain installation disturbs the soil to a degree, depending on its sensitivity and macrofabric (Rowe, 1968).

The effect of soil disturbance was included in the analyses by Barron (1948) and Hansbo (1979, 1981) by assuming an annulus of smeared clay around the drain. Within this annulus of diameter, d_s, (Figure 8), the remoulded soil has a lower coefficient of permeability, k_r, than the k_h of the undisturbed clay. This leads to a new boundary condition between the undisturbed zone and the smeared annulus, and it affects the solution by changing the factor F(n), which now becomes:

$$F_s(n) = \log_e\left[\frac{n}{s}\right] - 0.75 + \left[\frac{k_h}{k_r}\right]\log_e(s) \tag{6}$$

where s = smear zone ratio d_s/d_w

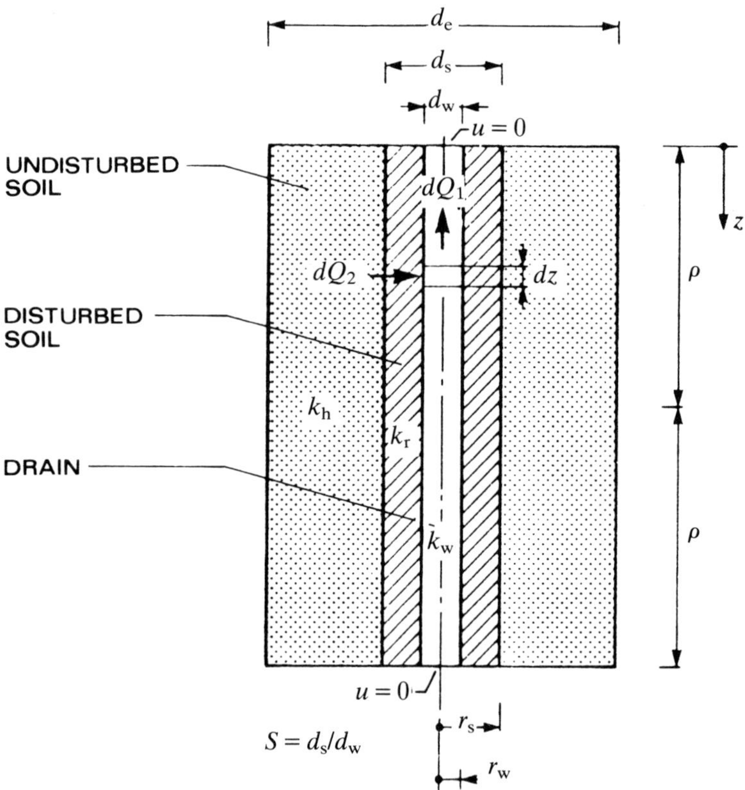

Figure 8 A vertical drain, including well resistance

The two added parameters, the ratio $s = d_s/d_w$ and the ratio k_h/k_r, are both difficult to estimate.

For non-displacement (jetted) sand drains, these ratios are usually assumed to be about equal to unity. For prefabricated low-displacement drains, Hansbo (1981) used a value of $s = 1.5$. Based on a theoretical analysis of the cone penetration process by Levadoux and Baligh (1980), this value corresponds to a shear strain of 15–50%. In any event, the problem is very difficult to analyse because, in practice, the amount of disturbance is related to the dimensions and shape of the driving 'shoe' at the mandrel tip (see also Chapter 2, Section 2.4 and Chapter 5, Section 5.15).

The ratio k_h/k_r could also be obtained experimentally from appropriate laboratory tests. Table 11 gives experimental results of the ratio k_v/k_r for some typical Italian soft clays. They are conservative ratios, because k_r was determined on completely remoulded specimens. In the field, the remoulding effect gradually decreases from the drain surface outward (see Chapter 5, Section 5.15).

Relevant in this context is the analysis reported by Jamiolkowski and Lancellotta (1984) and Jamiolkowski *et al.* (1985) of the settlement records from a trial preloading embankment on four types of drains at the Porto Tolle site (Figure 9). Figures 10a and 10b show the application of the Asaoka (1978) method (see Chapter 5, Section 5.9) to data from sections with jetted sand drains and Geodrains, respectively. The regression lines obtained for the four different drain types are compared in Figure 11, and the deduced c_h values are given in Figure 12. If we assume that the same c_h holds for all four areas, the k_h/k_r ratios range from 1.5 to 2 in order to obtain the same c_h

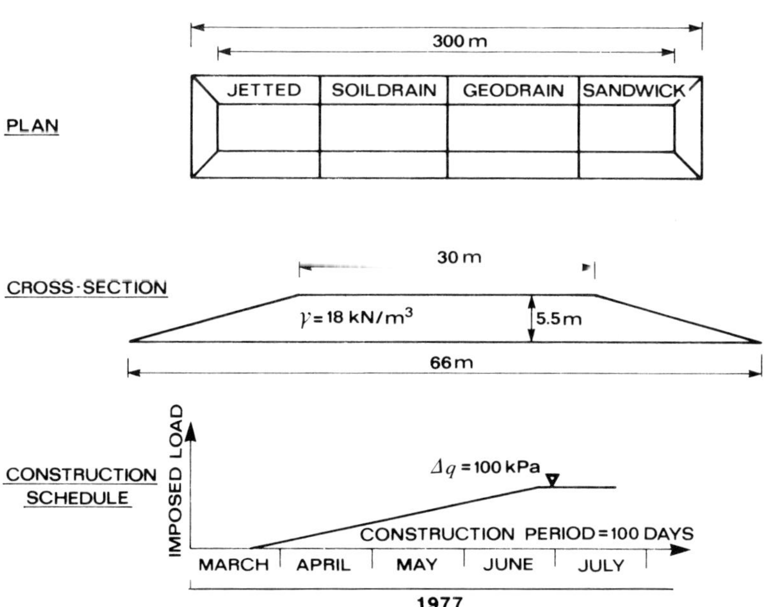

Figure 9 Geometry and loading of the trial embankment at Porto Tolle (Jamiolkowski and Lancellotta, 1984)

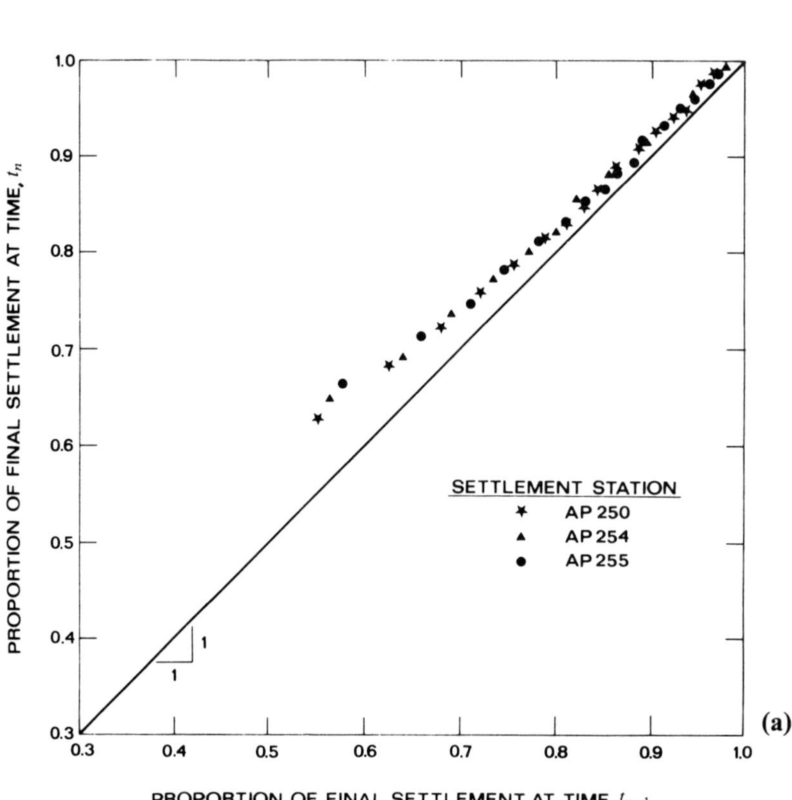

Figure 10 Asaoka's procedure for determining the coefficient of consolidation for horizontal flow of Porto Tolle silty clay below a preloading embankment: (a) jetted area; (b) Geodrain area (Asaoka, 1978; Jamiolkowski *et al.*, 1985)

value as for the jetted sand drains, i.e. this remoulding ratio is well below the values obtained from laboratory tests on completely remoulded specimens in Table 11.

Figure 13 shows that smear effects may be important both for the design of vertical drains and the analysis of their performance, difficult though it is to predict these effects.

The following points are of practical interest.

(1) The influence of smear increases with increasing drain diameter, d_w, because the ratio of $[F_s(n) - F(n)]$ to $F(n)$ increases as n decreases (Equations (4) and (6) and $n = d_e/d_w$).

(2) The degree of disturbance associated with these drains is appreciably lower than with conventional displacement sand drains because of the smaller volume of prefabricated drains relative to the volume of the treated soil.

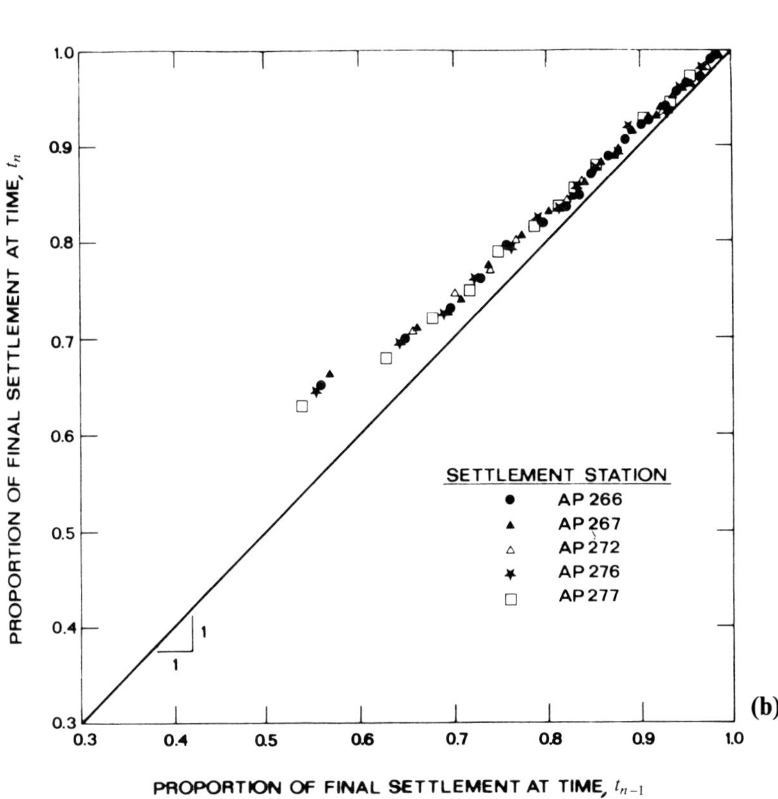

(b)

(3) When smear is present, a decrease in drain spacing is less effective in reducing consolidation times than for ideal drains, because the difference $[F_s(n) - F(n)]$ increases as n decreases.

3.4 Well resistance

As can be seen in Table 8 (page 11) most of the currently used prefabricated drains are band-shaped. They generally consist of a flat plastic core wrapped with a filter sleeve made of paper or non-woven geotextile filter fabric. The relevant features for the design and performance of these drains are their hydraulic properties: the discharge capacity, q_w, of the cross-section, and the filter permeability. The experimental assessment of these parameters is covered in Chapter 6, Section 6.5. The theoretical and practical relevance of well resistance which affects the discharge capacity is discussed below.

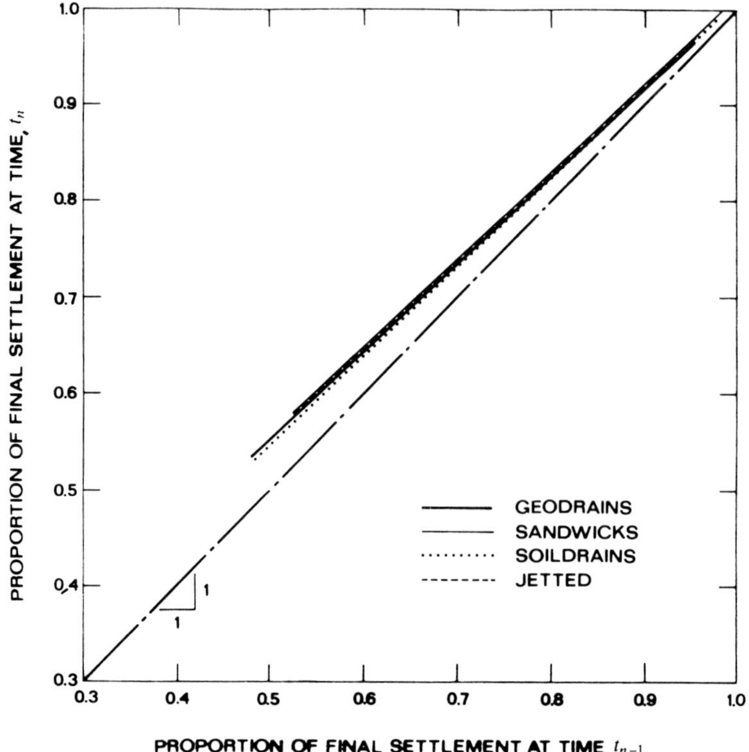

Figure 11 Comparison between rate of consolidation of a preloading embankment on four types of drains on Porto Tolle silty clay (Jamiolkowski *et al.*, 1985)

Table 11 Conservative values of the ratio of coefficient of permeability for vertical flow/coefficient of permeability of remoulded soil for soft clays in the normally consolidated range

Site	*Plasticity index*	*Coefficient of permeability for vertical flow,* k_v (m/year)	k_v/k_r
Panigaglia	45–65	2×10^{-3}	1.5–2
Porto Tolle	30	2×10^{-2}	4–8
Trieste	47 ± 10	3×10^{-3}	2–3

If, during consolidation, the discharge capacity of the drain is reached, the overall consolidation process is retarded. Barron (1948) first developed a solution for this case for equal strain conditions. For the same boundary conditions, Hansbo (1981) presented a similar solution, which showed that the degree of consolidation, U_{hz}, at depth, z, is still given by an equation of the same form as Equation (5), the factor $F(n)$ being replaced by:

$$F_w(n) = \log_e(n) - 0.75 + \pi z(2\ell - z)\left[\frac{k_h}{q_w}\right] \tag{7}$$

where ℓ = characteristic length of the drain, which is equal to:
half the drain length for fully penetrating drains, the so-called 'open-end' drains, or
entire length for partially penetrating drains, the so-called 'closed-end' drains
q_w = well discharge capacity $= A_w k_w$
A_w = cross-sectional area of the drain
k_h = coefficient of permeability for horizontal flow
z = depth

Aboshi and Yoshikuni (1967) gave numerical solutions obtained by using an electrical lumped-parameter model, and Yoshikuni and Nakanodo (1974) derived rigorous solutions for the more consistent free-strain case.

When assessing well resistance characteristics, there are three physical constraints which influence the use of the above solutions, as follows.

(1) The introduction of a finite well permeability affects the degree of consolidation, U_{hz}, which, as shown in Figure 14, is no longer constant with depth.
(2) The pore-water pressure in drain wells takes time to dissipate. Thus, the consolidation is particularly retarded in the lower part of the drain well.
(3) As expected from an inspection of Equation (7), the influence of a finite drain permeability on the consolidation rate increases as the drain length increases. This is shown in Figure 15, where the selected values of n are representative of the range of typical prefabricated-drain spacings.

Three parameters have been proposed to characterise well resistance.

3.4.1 Well-resistance parameter, W_R

The boundary condition at the drain interface can be expressed as follows (Hansbo, 1981).

$$\frac{\partial^2 u}{\partial Z^2} + 2\left[\frac{k_h}{k_w}\right]\left[\frac{\ell}{r_w}\right]^2 \frac{\partial u}{\partial \rho} = 0 \tag{8}$$

where $\rho = r/r_w$ and $Z = z/\ell$ are the normalised parameters

The second term of Equation (8) represents the effect of well resistance. According to Orleach (1983), it can be neglected if the parameter, W_R, is decreasing in magnitude,

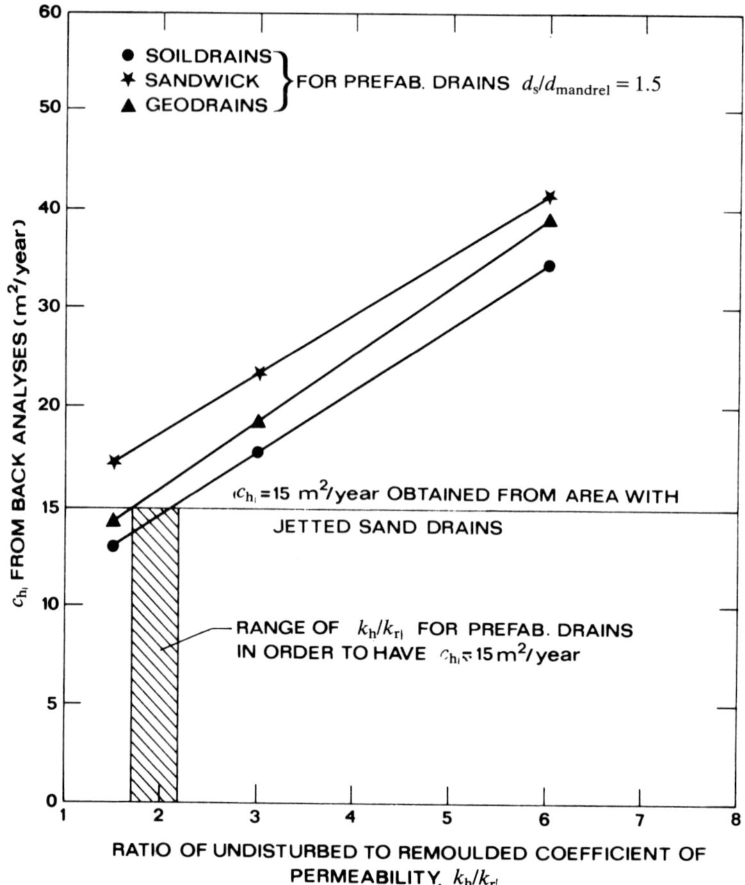

Figure 12 Values of the coefficient of consolidation for horizontal flow deduced from the regression lines in Figure 11 (Jamiolkowski *et al.*, 1985)

where

$$W_R = 2\left[\frac{k_h}{k_w}\right]\left[\frac{\ell}{r_w}\right]^2 = 2\pi\left[\frac{k_h}{q_w}\right]\ell^2 \tag{9}$$

3.4.2 Well-resistance parameter, *L*

Yoshikuni and Nakanodo (1974) used the parameter, L, to represent the effect of well resistance,

where

$$L = \frac{32}{\pi^2}\left[\frac{k_h}{k_w}\right]\left[\frac{\ell}{d_w}\right]^2 = \frac{4W_R}{\pi^2} \tag{10}$$

They show that the well-resistance effect is less than 10% when $L < 0.25$ (at $\bar{U}_h = 80\%$).

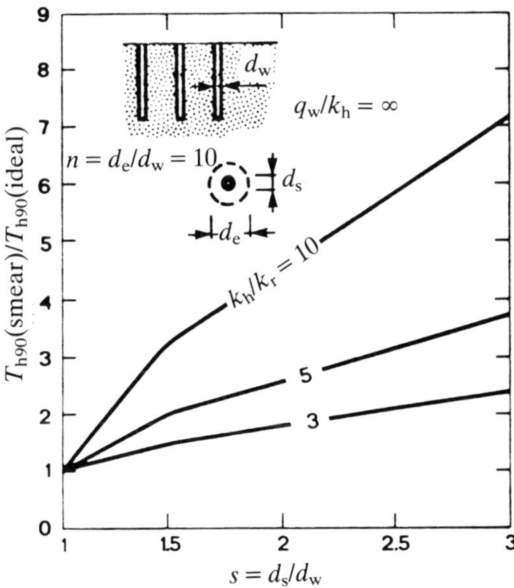

Figure 13 The effect of smear on consolidation rate (after Lancellotta, Maniscalco and Battaglio, 1981)

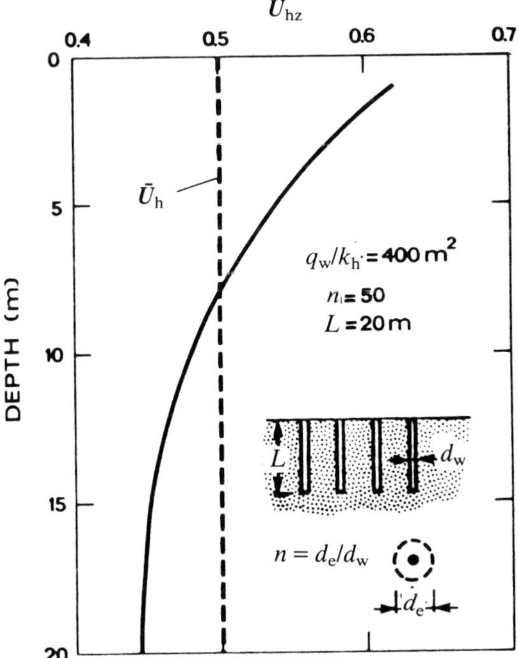

Figure 14 An example of the variation of degree of consolidation with depth for drains with well resistance (after Lancellotta, Maniscalco and Battaglio, 1981)

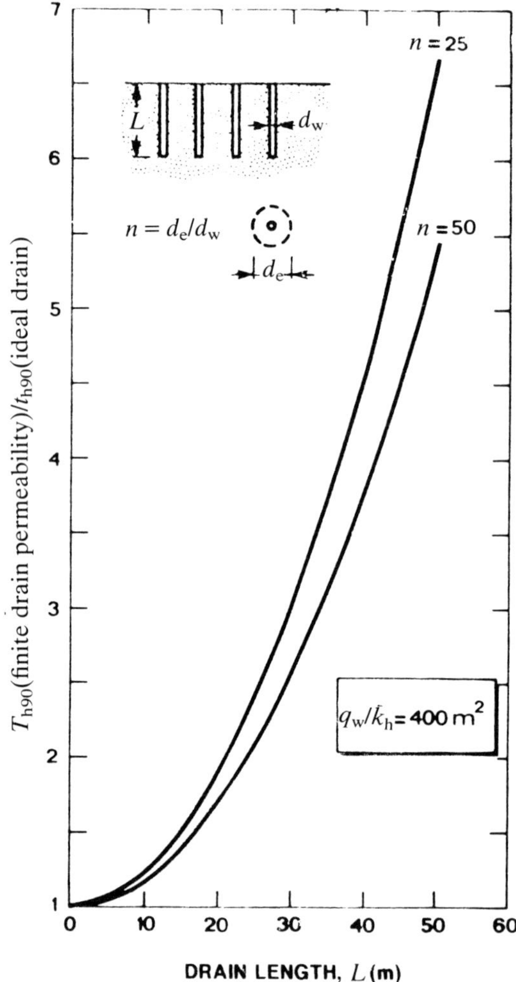

Figure 15 The influence of drain length on consolidation rate (Lancellotta, Maniscalco and Battaglio, 1981)

3.4.3 Well-resistance parameter, *R*

Aboshi and Yoshikuni (1967) suggested that the parameter

$$R = \frac{(n^2-1)}{n^2\,F(n)}\left[\frac{k_h}{k_w}\right]\left[\frac{\ell}{d_w}\right]^2 \tag{11}$$

could be used. In this case also, the well-resistance effect is less than 10% when $R < 0.02$.

3.4.4 Typical discharge capacities

As shown in Chapter 6, Section 6.5, typical values of the discharge capacity of many prefabricated drains are around 500 down to 100 m^3/year. Using these values with the permeability and geometrical parameters in Table 12, it is possible to show that, for most of the time, the well resistance does not affect the design. But when using very long drains under high lateral stresses, the increase in the time to reach a degree of consolidation of 80% (expressed in Table 12 as ΔT_{80}) may be of the order of 30% or more.

3.4.5 Increase of well resistance

The conclusion in Section 3.4.4 is valid from a theoretical point of view, but, in practice, well resistance can develop and increase for three reasons.

(1) The deterioration of the drain filter may lead to a significant reduction of the cross-section.
(2) Fine soil particles may pass through the filter and decrease the area available for flow (siltation).
(3) Folding of the drain because of large settlements may result in a reduced discharge capacity.

All these aspects are still difficult to quantify (see Chapter 6, Section 6.5).

3.5 Combined radial and vertical drainage

The effect of vertical drainage on the consolidation rate was not taken into account in the Barron (1948) equations. However, Orleach (1983) used Carrillo's (1942) relation to account for combined radial and vertical drainage, and the results are shown in Figures 16 and 17. The effect of vertical drainage is predominant near the top

Table 12 Examples of the increase in consolidation times ΔT_{80} for typical values of drain spacing ratio, *n*, drain length, ℓ, and discharge capacity, q_w (after Aboshi and Yoshikuni, 1967; Yoshikuni and Nakanodo, 1974)

Drain spacing ratio	*Drain length* (closed end)	*Discharge capacity*	*Coefficient of permeability of soil*	*Well resistance parameters and increase in consolidation time*			
n	ℓ *(m)*	q_w *(m³/year)*	*k* *(m/s)*	L^*	ΔT_{80} *(%)*	$R^\dagger$	ΔT_{80} *(%)*
25	15	100	10^{-9}	0.18	11	0.023	10
25	15	500	10^{-9}	0.036	N	N	N
25	25	100	10^{-9}	0.50	22	0.064	30
25	25	500	10^{-9}	0.10	11	0.013	5

N negligible.
*Yoshikuni and Nakanodo (1974).
†Aboshi and Yoshikuni (1967).

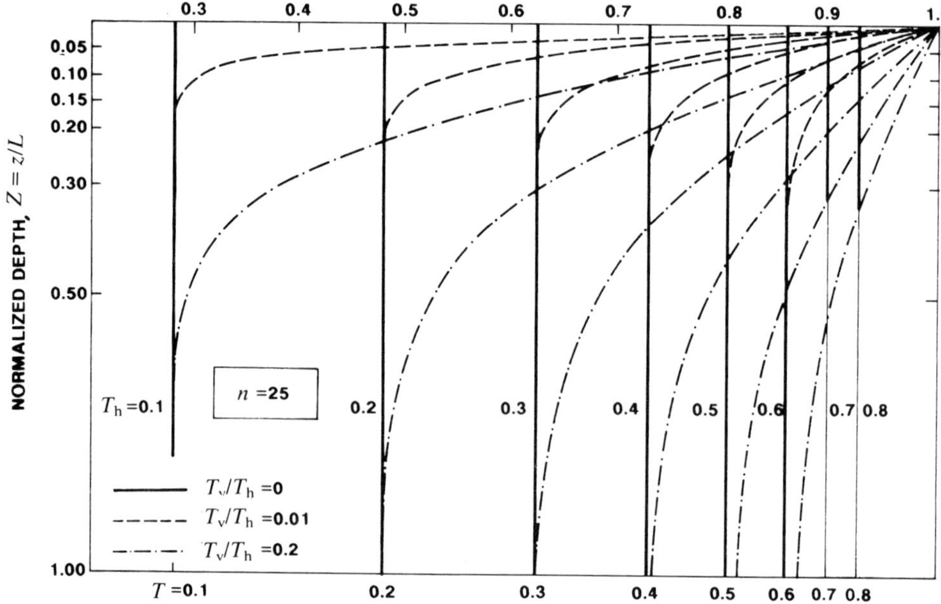

Figure 16 The effect of vertical drainage on the average degree of consolidation versus depth for different time factor ratios and a drain spacing ratio of 25 (after Orleach, 1983)

permeable boundary, with a zone of influence increasing with time and with the ratio T_v/T_h (Figure 16).

This fact may be relevant when analysing data from piezometers in order to obtain the coefficient of consolidation. For ratios of T_v/T_h lower than 0.01, the zone of influence is about 0.2–0.3 times the non-dimensional depth, Z. This leads to an overestimate of c_h if the vertical drainage is neglected when using piezometer data from within this depth. The same conclusion applies to settlement data, because, from Figure 17, it may be seen that the vertical drainage should be considered when the ratio T_v/T_h is greater than about 0.01–0.02.

3.6 Time-dependent loading

In practice, the full load cannot be applied instantaneously, so that, in effect, a gradually increasing, or ramp-type, loading occurs.

Often (and particularly when vertical drains are used) the consolidation during this stage may be a significant part of the entire process. Thus, it cannot be neglected.

Taylor (1948) first suggested an approximate method to account for consolidation during construction without drains. He considered the linear ramp loading problem to be equivalent to that of an instantaneous load applied at the middle of the loading period. This procedure has also been used in practice with vertical drains, and exact solutions have been developed by Schiffman (1958) and Olson (1977).

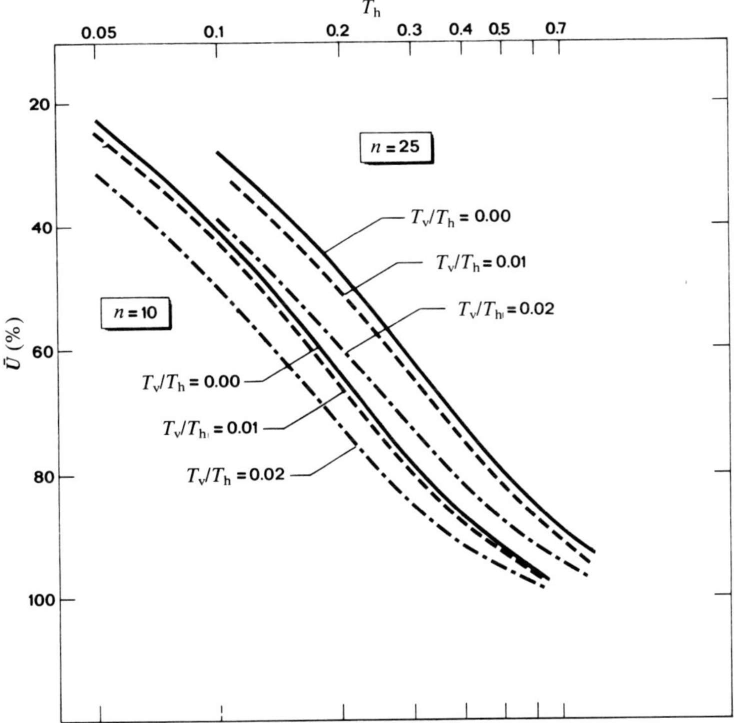

Figure 17 The average degree of consolidation/log time for different time factor ratios of horizontal/vertical ratios for drain spacing ratios of 10 and 25 (Orleach, 1983)

Olson's (1977) equations for the excess pore pressure, u, and the average degree of consolidation, $\bar{U}$, are:

(a) $T_h \leqslant T_{hc}$

$$u = \frac{B}{A} \cdot \frac{u_o}{T_{hc}} [1 - \exp(-A\,T_h)]$$

$$\bar{U}_h = \frac{1}{T_{hc}} \left\{ T_h - \frac{1}{A} \left[1 - \exp\left(-A\,T_h \right) \right] \right\}$$

(b) $T_h \geqslant T_{hc}$

$$u = \frac{B}{A} \cdot \frac{u_o}{T_{hc}} [\exp(A\,T_{hc}) - 1]\ \exp(-A T_h)$$

$$\bar{U}_h = \frac{1}{A T_{hc}} [\exp(A T_{hc}) - 1]\ \exp(-A T_{hc})$$

where $A = 2/\mathrm{F}(n)$

$$B = \frac{1}{n^2\mathrm{F}(n)} \left[n^2 \log_e \left[\frac{r}{r_w} \right] - \frac{(r^2 - r_w^2)}{2 r_w^2} \right]$$

$$T_h = (c_h\, t)/r_e^2 \tag{11}$$

The main problem with these equations arises if the applied load stresses an overconsolidated clay stratum into the normally consolidated range. In this case, the coefficient of consolidation decreases significantly during stage loading, so the above equations no longer apply.

It has to be mentioned that Schiffman's (1958) solution also treats the case of permeability being linearly dependent on the excess pore pressure. Unfortunately, such a relationship does not appear to apply to natural clays.

In fact, comprehensive laboratory research on the permeability of natural clays by Tavenas, Tremblay and Leroueil (1983) leads to the following conclusions.

(1) None of the relations proposed in the literature between permeability and void ratio is generally valid, irrespective of the plasticity and fabric of the clay, its initial void ratio, or the range of void ratio change.
(2) Among the existing relations, that proposed by Taylor (1948):

$$\Delta e = C_k \Delta(\log k) \tag{12}$$

where C_k is an empirical permeability change index (Tavenas, Jean, Lablond and Leroueil, 1983).

gives the best interpretation of experimental data for initial void ratios less than 2.5 and for volumetric strains of practical interest. On the basis of this latter conclusion, the relation between the pore-pressure change and the permeability should be:

$$(u_o/u) = \left[\frac{k}{k_o} \right]^{C_k/C_c}$$

where C_c = compression index
k_o = initial coefficient of permeability

This is linear if $C_k/C_c = 1$, i.e. only when the coefficient of consolidation is constant during the process. Therefore, to account for the dependence of the soil parameters on the stress history, a numerical treatment of the problem is needed (see Section 3.7).

Stage-construction procedures, used if the total load cannot be applied in one step without stability problems, are obviously time-dependent. This case usually is solved by superimposing the effects of each single-stage load. Again, this simple procedure may be questioned if the soil characteristics change during the consolidation process.

However, when using stage-loading procedures, the main problem (particularly with lightly overconsolidated clay) is to evaluate the soil's strength increase after each load increment. This aspect is beyond the scope of this Report and the reader should refer to Ladd (1976, 1986, 1989).

3.7 Limitations of the Terzaghi–Barron theory

The analytical solutions developed by Terzaghi (1923) and Barron (1948) are restricted to the linear–elastic behaviour of clay, a simple sequence of loading, small deformations, and no spatial variations of soil properties. Much effort has been made to overcome most of these limitations, but this work has been directed to the simple case of consolidation with vertical flow (Taylor, 1942; Gibson and Lo, 1961; Barden, 1965; Davis and Raymond, 1965; Janbu 1965; Mikasa, 1965; Gibson, England and Hussey, 1967; Gibson, Schiffman and Pu, 1967; Raymond, 1969; Suklje, 1969; Davis, 1971; Garlanger, 1972; Viggiani, 1973; Mesri and Rokhsar, 1974; de Simone and Viggiani, 1976; Olson and Ladd, 1979; Burghignoli, 1979).

However, because the conclusions of these analyses should also be valid for the case of radial flow, they are summarised below.

(1) The case of non-linear soil behaviour, having c_v constant, but permeability and compressibility characteristics changing with vertical effective stress, was analysed first by Davis and Raymond (1965). Because of soil non-linearity, the average degree of consolidation in terms of settlement, $\bar{U}_s$, does not equal that in terms of pore pressure, $\bar{U}_p$. While the former ($\bar{U}_s$) is consistent with that of Terzaghi's theory, the rate of decay of the excess pore pressure is inversely proportional to the ratio σ'_f/σ'_{vo}

(2) The influence of the stress history is shown by the fact that both the average consolidation ratios $\bar{U}_s$ and $\bar{U}_p$ increase as the ratio σ'_p/σ'_{vo} increases, for a given ratio of C_r/C_c (Mesri and Rokhsar, 1974).

(3) Non-linear soil behaviour can also explain the abnormal pore-pressure decay observed in highly structured clay deposits (Figure 18). This has been shown by Mesri (1979) and Mesri and Choi (1979), who analysed the Gloucester test fill (Crawford and Burn, 1976) using a linear e–log k relationship and accounting for the highly non-linear void ratio–effective stress relationship. The results in Figure 18 show:
   (a) a rapid pore-pressure dissipation with little settlement during the early part of the consolidation along the flat recompression curve;
   (b) a substantial settlement with relatively little pore-pressure dissipation when the vertical effective stress exceeds σ'_p and collapse of the clay structure occurs.

(4) Weber (1969) reports a case of an embankment over peat that exhibited a settlement equal to 80% of the thickness of the consolidating stratum. In such cases, the Terzaghi hypothesis of small strains no longer applies. To account for large strains, a numerical treatment of the problem is usually made. Olson and Ladd (1979) showed that the assumption of small strains introduces an error of

8% in terms of $\bar{U}_p$ up to strains of 30% (see also Mesri and Rokhsar, 1974). The increase in consolidation rate may be more significant for larger strains. This effect must be taken into account.

(5) It is more difficult to give definitive conclusions about the effects of secondary compression. Theoretically, its influence increases (i.e. a slower rate of consolidation occurs) as the ratio C_α/C_c increases or, for a given C_α/C_c ratio, as the ratio σ'_f/σ'_{vo} decreases. The precise amount of its influence depends on the specific rheologic model and the input parameters used.

But the most important question is whether secondary compression is a separate phenomenon while excess pore pressure dissipates during primary

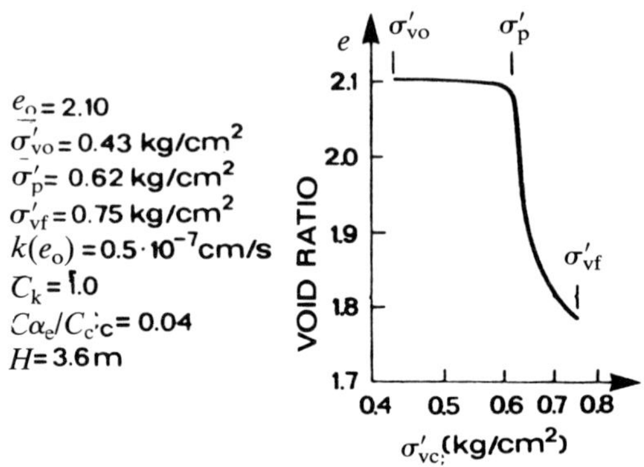

(a) INPUT DATA FOR THE THEORETICAL ANALYSES

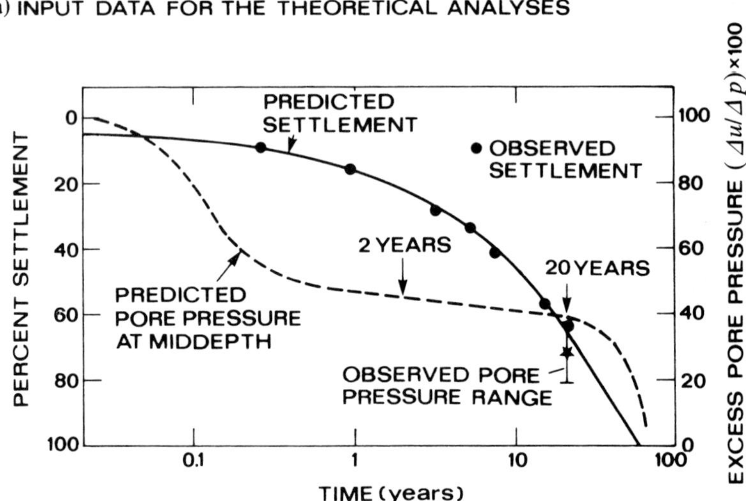

(b) OBSERVED AND PREDICTED SETTLEMENT AND PORE PRESSURE BEHAVIOUR

Figure 18 Consolidation analyses of the Gloucester, Canada test fill (after Mesri, 1979)

consolidation. As pointed out by Ladd *et al.* (1977) and Jamiolkowski *et al.* (1985), two hypotheses have been proposed to describe this phenomenon (Figure 19). Hypothesis (A) assumes that secondary compression occurs only after the end of primary consolidation, and it predicts that sample thickness (i.e. drainage height, H_d, and, hence, the time required for pore-pressure dissipation) has no effect on the location of the end of primary compression curve.

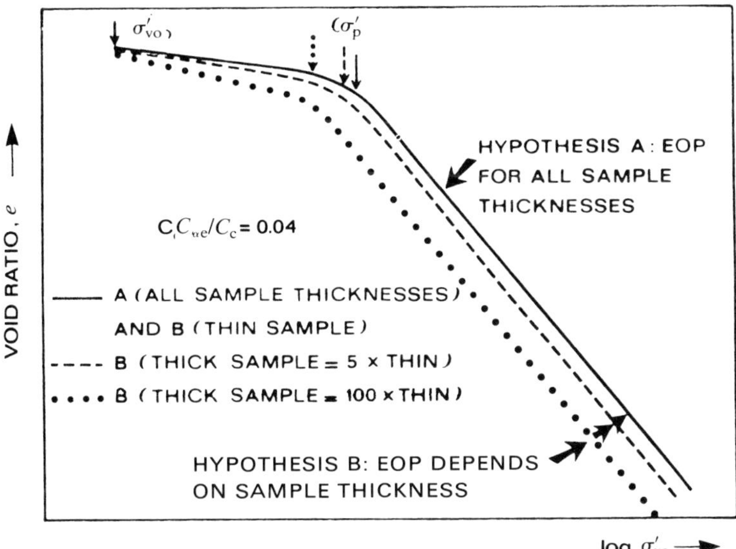

(b) STRAIN VS TIME FOR OCR = 1 SAMPLES HAVING EQUAL INITIAL CONDITIONS AND $\Delta\sigma$

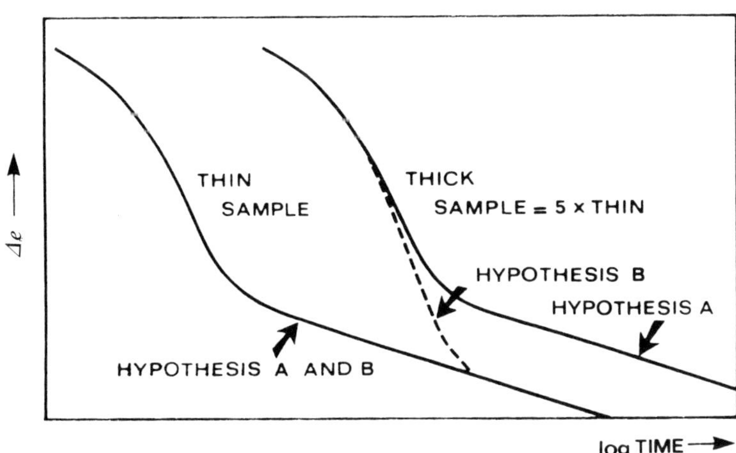

Figure 19 An illustration of hypotheses (A) and (B) in terms of: (a) stress versus strain; and (b) strain versus time

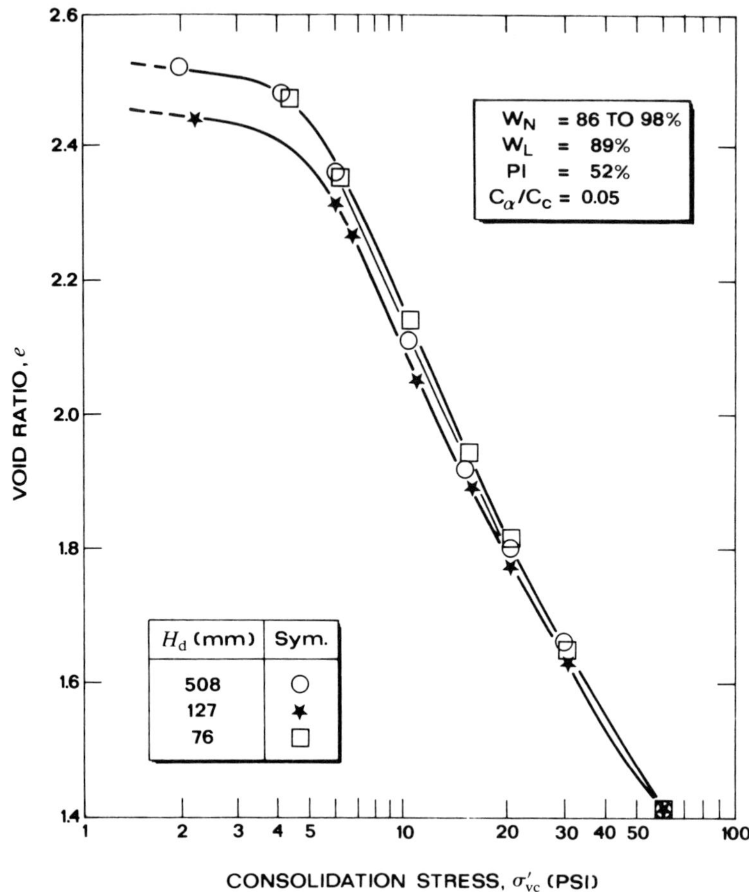

Figure 20 End of primary compression curves for San Francisco Bay mud with varying drainage heights (Mesri and Choi, 1985)

Hypothesis (B) (Figure 19) assumes that some sort of 'structural viscosity' is responsible for secondary compression, that this phenomenon occurs during pore-pressure dissipation, and therefore that the strain at the end of primary consolidation increases with sample thickness. Mesri and Choi (1985) reviewed the previous attempts to investigate this phenomenon, and they present strong evidence from an 8-year-long test programme on three natural clays. The data shown in Figure 20 are from three oedometer tests on specimens ranging from 76 to 508 mm in height, and all gave a unique end of primary consolidation curve. These results strongly support hypothesis (A), namely that creep occurs only after dissipation of excess pore pressure. Nevertheless, according to Mesri (1985), hypothesis (A) has to be considered an empirical criterion, being based on observed behaviour of soils but not the result of a rigorous analysis. In greater detail (Mesri and Choi, 1985), the change in void ratio Δe at time, t, should be expressed as in Equation (13) below.

$$\Delta e = \int_0^{t_p} \left[\left(\frac{\partial e}{\partial \sigma'} \right)_t \frac{d\sigma'}{dt} + \left(\frac{\partial e}{\partial t} \right)_{\sigma'} \right] dt + \int_{t_p}^{t} \left(\frac{\partial e}{\partial t} \right)_{\sigma'} dt \tag{13}$$

where t_p is the time to the end of primary consolidation

This implies that the time-dependent compressibility, $(\partial e/\partial t)_{\sigma'}$, and the stress-dependent compressibility, $(\partial e/\partial \sigma')_t$, both contribute to the change in void ratio during the primary consolidation. The above-mentioned working hypothesis comes from the fact that at present it is not possible to separate $(\partial e/\partial \sigma')_t$ from $(\partial e/\partial t')_{\sigma'}$ when the effective stress is changing with time, i.e. $\partial \sigma'/\partial t \neq 0$.

4 Information needed for design

The theme of the Eighth European Conference on Soil Mechanics and Foundation Engineering held in Helsinki in 1983 was *Improvement of Ground.* From the papers, the state-of-the-art report, and the discussions, there appears to be almost universal agreement on the need for comprehensive investigation, improved predictive capability and standardisation of materials and testing. These are discussed in the following sections.

4.1 Comprehensive investigation

A successful application of precompression techniques to speed up consolidation of soft clay deposits requires a properly programmed and comprehensive *in-situ* and laboratory soils investigation. Such an investigation should provide the following information.

(1) Stress history of the deposit.
(2) Macrofabric, geometry of the drainage paths, and the drainage boundaries.
(3) Consolidation and permeability characteristics in both vertical and horizontal directions.
(4) Stress–strain and strength characteristics.
(5) Evaluation of possible changes to the stress–strain, strength and water-flow characteristics of the soil caused by the proposed method of drain installation.

The first two items are obviously essential for the prediction of the magnitude and rate of settlements. The stress–strain and strength properties are required because a rational design of vertical-drain systems and preloading fills also involves analysing the stability of the fill during, and at the end of, construction. Frequently, because of the low initial undrained strength of a soft clay, the preloading has to be placed in two or more stages in order to avoid foundation failures. This requires the stability of the preloading embankment to be verified under so-called 'partially consolidated' conditions, which, in turn, requires an evaluation of the increase of the shear strength of the foundation soils with consolidation.

The importance of stress history is frequently not fully recognised by designers; its impact on precompression design is twofold: (i) the relationship between stress history and the magnitude of total stress imposed on the soil governs whether there is a real need for these techniques, and (ii) there is a strong dependence of c_h and c_v on the stress history of the deposit, as shown in Figure 21. This dependence needs to be considered in the design if a reliable estimate of the settlement rates of overconsolidated deposits is desired.

These points are discussed in detail in Chapter 5.

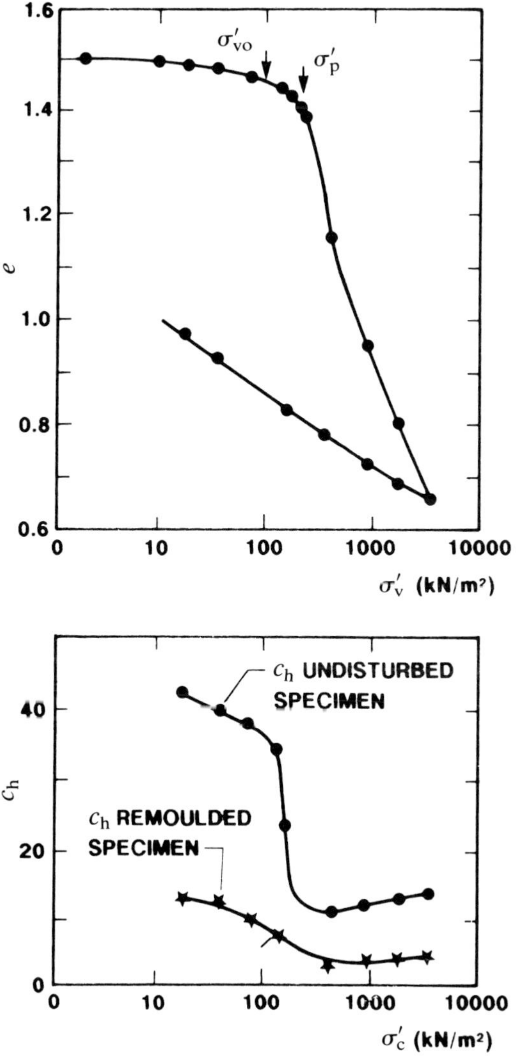

Figure 21 The effects of stress history and soil disturbance on the coefficient of consolidation (Jamiolkowski, Lancellotta and Wolski, 1983a)

4.2 Improving predictive capability

It is widely recognised that some of the most difficult Class A predictions (Lambe, 1973) are those for settlement rates and pore-pressure dissipation in the field. The main reason for this is the difficulty of making a reliable evaluation of the average or mass flow and consolidation properties of natural soil deposits.

In order to assist the designer in the proper selection of the experimental procedures leading to a more reliable prediction of the settlement and consolidation rates, the recommendations given in Table 13 have been developed. They represent a compromise between laboratory and *in-situ* testing techniques, e.g. in using laboratory determinations of m_v, the coefficient of volume compressibility, with field determinations of the coefficients of permeability. The advantages and limitations of each have been discussed by Jamiolkowski, Lancellotta and Wolski (1983a), who also discuss the uncertainties involved in the evaluation of c_h (and c_v) from pore-pressure and settlement data obtained from instrumented trial embankments. These points are summarised in Chapter 5.

Table 13 Procedures necessary to improve the reliability of predictions of the settlement and consolidation rates of vertical-drain installations

For all projects: Investigate the macrofabric and drainage paths of the deposit by means of borings (preferably with continuous undisturbed sampling) and piezocone testing

When dealing with deposits of:

	(1) Uniform clay	*(2) Clay with macrofabric*
For small projects	Laboratory c_h Piezocone dissipation tests	Field k_h + laboratory m_v Piezocone dissipation tests
For large projects	Field k_h + laboratory m_v Piezocone dissipation tests Trial embankment*	Trial embankment Field k_h + laboratory m_v Piezocone dissipation tests

*May not be strictly necessary.

4.3 Material standardisation and testing

It is widely recognised that standardisation of material requirements and testing procedures for the characterisation of prefabricated vertical drains is needed urgently. In view of the proliferation of new types of prefabricated drains and their high rates of installation, millions of metres of drains might be installed which may not function properly or may deteriorate rapidly after installation without such standards. A number of factors leading to improved testing and specifications of prefabricated drains are discussed in Chapter 6.

The information needed for design relates to the ground and to the drainage system to be installed. The relevant geotechnical properties and characteristics required for the design of vertical-drain systems are discussed in some detail in Chapter 5. Topics such as the consolidation and permeability characteristics, the stress history and the macrofabric of the deposit, its undrained shear strength, secondary compression properties, and the effects of drain installation and soil disturbances on design are also included. Chapter 6 is concerned with the physical and mechanical characteristics of the drains themselves and how these factors impact on design and performance; discharge capacity, filter permeability, drain geometry and mechanical properties and durability of the drain materials are also included.

5 Geotechnical characteristics

5.1 Geotechnical information needed to design vertical-drain systems

The specific parameters governing the design and performance of precompression techniques with vertical drains in soft cohesive soil deposits are the following.

(1) Stress history of the deposit, σ'_p, and overconsolidation ratio (OCR).
(2) Consolidation coefficients for horizontal, c_h, and vertical, c_v, flow.
(3) Extent of the smeared zone, d_s, and other drain installation effects.
(4) Coefficient of permeability of the remoulded clay within the smeared zone, k_r.
(5) Undrained shear strength, c_u, and undrained Young's modulus, E_u.
(6) Secondary compression index, C_α, or secondary compression ratio, $C_{\alpha\varepsilon}$.
(7) Discharge capacity of the drain, q_w, and its variation with total lateral stress, σ_h, and time, t.

All these parameters, except the discharge capacity, are geotechnical in nature, i.e. they depend on the specific geotechnical conditions at the site. Discharge capacity is a function of the drain, and it is discussed in Chapter 6, Section 6.5.

In addition to the above, other geotechnical considerations of the site which also have to be addressed at least qualitatively in design include the macrofabric of the deposit, drainage-boundary characteristics, and the effect of the size of the loaded area relative to the thickness of the compressible layer, i.e. one- and three-dimensional loading.

All the geotechnical information can be obtained by conventional site investigation techniques, including laboratory tests on undisturbed samples, as well as by a number of specialised *in-situ* tests. Another approach, very useful for major projects at sites with significant macrofabric, is to back-calculate the parameters from the results of field measurements on full-scale trial embankments or other prototype structures (Table 13). The appropriate use of these techniques is discussed in the remainder of this chapter.

General aspects of site investigation, subsurface exploration, and sampling are covered in detail by Weltman and Head (1983). Conventional laboratory testing procedures are discussed by Lambe (1951), Bowles (1978) and Head (1982), among others. For a comprehensive summary of laboratory and field permeability measurements, see Olson and Daniel (1979).

National standards organisations issue standard methods for some *in-situ* and laboratory tests, but the forms of *in-situ* testing to determine the soil properties for the design of preloading and deep-drainage projects are developing very rapidly. The reports by Jamiolkowski, Lancellotta and Wolski (1983a) and Jamiolkowski *et al.* (1985) remain basic references for Sections 5.2, 5.3, 5.5 and 5.6.

5.2 Consolidation and permeability characteristics

Even with all the progress achieved in experimental soil mechanics during the last two decades, a reliable prediction of the consolidation rates of soil deposits is still one of the most difficult tasks for the designer. As noted by Leminen and Rathmayer (1983) and Hansbo (1983c), accurate determination of the consolidation and permeability characteristics has important economic consequences. The problem of reliable predictions is not so much because of imperfections in the theories used, but rather from the difficulties and uncertainties involved in the experimental determination of the permeability and consolidation characteristics at a site. Existing experience with the determination and selection of design values of c_h and k_h is summarised in Sections 5.3 and 5.4.

5.3 Consolidation and permeability characteristics: laboratory determination

The usual approach for the determination of c_h and k_h in the laboratory is from ordinary oedometer tests (incremental or continuous-loading) with vertical drainage. Then, based on local experience or on a knowledge of the stratigraphy, c_h (or k_h) is assumed to be either equal to, or somewhat larger than, c_v (or k_v). A better approach is to determine c_h (and k_h) in oedometer tests in which radial or horizontal drainage is allowed. Such tests can be easily carried out in Rowe cells (Rowe and Barden, 1966). Alternatively, vertical (rather than the conventional horizontal) oedometer specimens could be cut from undisturbed tube samples and tested in the usual manner. Although the laboratory strain conditions differ from those in the field, this procedure has been used successfully a number of times in Finland (Leminen and Rathmayer, 1983).

It is also possible to perform consolidation tests in a triaxial cell during which simultaneous radial and vertical pore-water flow can take place. If only radial flow is desired, the caps and bases of the triaxial cell require some modification. In order to model field conditions more accurately, no lateral strain should be permitted. Special radial-flow permeameters and consolidation cells specifically designed to measure c_h and k_h have been described by Olson and Daniel (1979), Collins (1979), Trautwein (1980), and Trautwein, Olson and Thomas (1981). Although only used for research so far, the devices appear to be suitable for use in practice, as a complement to the Rowe cell. Olson and Daniel (1979) and Tavenas, Tremblay and Leroueil (1983) describe standard and special tests, as well as the sources of error for permeability determinations. The use of soil parameters obtained from conventional tests on 76–102 mm diameter samples for the design of vertical drains may yield reasonable predictions of the consolidation rate, but only in those deposits in which macrofabric features are absent.

In soil deposits with a well-developed macrofabric, the use of flow and consolidation parameters based on conventional laboratory tests for vertical-drain design generally leads to an appreciable underestimate of the consolidation rate and, hence, to an uneconomical drain design (Rowe, 1968, 1972). In fact, the presence of thin layers and seams of more permeable material, as well as discontinuities resulting from desiccation and the decomposition of organic matter, require that the following points should be considered in design of the drains.

(1) The deposit may have a significant anisotropy with respect to permeability and coefficient of consolidation, i.e. $k_h > k_v$ and $c_h > c_v$.
(2) Laboratory-determined values of c_h (c_v) and k_h (k_v) depend on the dimensions of the test specimen.
(3) The possible disturbance of test specimens generally causes the laboratory c_h to be lower than the field c_h.

For design, Rowe (1968, 1972), McGown, Gabr and McNeil (1979) and McGown and Hughes (1981) recommend large-diameter (152–254 mm) Rowe cell tests on high-quality, large-diameter piston samples.

Additional comments on macrofabric are made in Section 5.11.

In the laboratory, oedometer tests run on high-quality undisturbed specimens of a typical soft clay usually show variations of the consolidation coefficient, c_h, with vertical effective stress, as in Figure 21. For comparison, the same illustration shows the value of c_h determined on a remoulded specimen of the same clay at the same water content.

Figure 21 also illustrates the following important trends which, according to Ladd (1971), are believed to be typical for the consolidation characteristics of normally consolidated, slightly overconsolidated, and aged natural clay deposits.

(1) c_h decreases significantly as the consolidation stress, σ'_{vc}, approaches σ'_p.
(2) c_h remains approximately constant or increases slightly in the normally consolidated range.
(3) Consequently, c_h on reloading is much larger than during initial loading.
(4) The c_h of remoulded material is lower than the c_h of undisturbed soil tested at the same stress level. The difference is particularly large in the overconsolidated range.

Figure 21 shows how pronounced may be the influence of partial disturbance of the test specimens on the values of c_h measured in the laboratory. This explains at least partly why many case records report an observed field consolidation rate which is much faster than the one computed on the basis of conventional laboratory test results.

On the other hand, there are case records which show excellent agreement between the observed field consolidation rates and those predicted on the basis of laboratory permeability and consolidation tests measured on small specimens. Such agreement appears to be possible in the case of homogeneous clay deposits without macrofabric. However, it should not be overlooked that other factors may, under certain circumstances, lead to a slow-down in field rates of pore-pressure dissipation. These factors, which are rarely considered because they are difficult to evaluate, include the following points.

(1) The large decrease of c_h and void ratio with increasing vertical effective stress σ'_v level.
(2) Collapse of soil structure when $\sigma'_v > \sigma'_p$.
(3) Shear strain or creep-generated excess pore pressure.

5.4 Consolidation and permeability characteristics: field determination

In view of the limitations of laboratory tests, many geotechnical engineers agree that the design of vertical drains should be based on permeability and consolidation characteristics obtained from appropriate *in-situ* tests. The conventional approach is to perform permeability tests in piezometers. A comprehensive review of the procedures involved and related interpretation methods has been given by Mitchell and Gardner (1975), Milligan (1975), and Olson and Daniel (1979). Tremblay (1983) and Jamiolkowski *et al.* (1985) summarise recently developed *in-situ* tests such as self-boring permeameters and piezocones.

5.4.1 Tests in piezometers

Tests in piezometers have been used, apparently successfully, for some 30 years to determine the *in-situ* k_h or c_h by both constant-head and falling-head tests. Usually, constant-head tests are preferred because test interpretation poses fewer uncertainties although, in general, interpretation is made more difficult because the imposed changes in effective stresses cause changes in the consolidation and flow parameters. In addition, because the *in-situ* tests are conducted prior to construction (i.e. at relatively low effective stress levels compared to those acting later under the embankment) a few companion laboratory tests should be carried out to assess the influence of an increase in effective stress on the measured properties (Rowe, 1968; McGown and Hughes, 1981).

According to Baligh and Levadoux (1980) the main problems with the determination of c_h and k_h in the field from tests using piezometers are as follows (Tremblay (1983) also considers the first two to be of most significance).

(1) Disturbance and remoulding effects from piezometer installation, particularly when piezometers are pushed in (as is common in soft deposits). This invariably causes changes in the stresses and properties such as the void ratio of the surrounding soil mass. As in the case of laboratory tests (Figure 21), the

correctness of the measured c_h and k_h values depends on the severity of these changes, and on the sensitivity of the soil.

(2) Smearing and clogging of the porous element of the piezometer.

(3) Danger of hydraulic fracturing if an excessive increase of pore pressures in the surrounding soil occurs, either during installation or as a result of the applied head during the test.

(4) The flexibility of the testing system and gas in the pore water, which cause time lag. Also included are problems related to instrument performance and testing procedures, factors to which the test is very sensitive.

(5) Elimination of macrostructure features such as fissures, varves or thin layers in the soil in the immediate vicinity of the piezometer.

(6) The effect (in both constant- and variable-head tests) of the soil surrounding the piezometer tip being subject to changes in effective stresses because of the imposed pore pressures. Thus, in a constant-head test where flow is directed from the piezometer towards the soil (an outflow test), the coefficients of permeability and consolidation represent unloading and reloading conditions. These values of k_h and c_h reflect the soil behaviour in the overconsolidated state, and so tend to be appreciably higher than those obtained under conditions of primary compression (where the vertical effective stress exceeds σ'_p). They are thus appropriate for field problems where $\sigma'_{vo} + \Delta\sigma < \sigma'_p$.

For the situation where $\sigma'_{vo} + \Delta\sigma > \sigma'_p$, it is preferable to carry out constant-head permeability tests where flow is directed towards the piezometer (inflow test). In this case, k_h or c_h are determined in the effective stress ranges related to primary compression.

However in 1984, Tavenas, in a personal communication to the Authors, disagreed with the idea that inflow tests are necessary to measure k in the normally consolidated range. A significant error in the computed permeability may result because the available theories do not properly account for the complex consolidation process involved in such tests. Tavenas suggests, on the basis of laboratory investigations (Tavenas, Jean, Lablond and Leroueil, 1983) that the value of k at the *in-situ* stress and void ratio together with the permeability change index, C_k, are sufficient to describe the permeability behaviour for loading situations. In this case, the *in-situ* permeability is best obtained from outflow tests which, if the initial void ratio, e_o, stays nearly constant and the clay structure can be maintained, will be reasonably accurate. Tavenas also suggests that the permeability change index, C_k, may be obtained from laboratory tests such as falling-head tests in the oedometer or simply from an empirical correlation $C_k = 0.5e_o$.

Variable-head tests in piezometers are much less desirable than constant-head tests (Tremblay, 1983). The changes in effective stress to which the soil around the piezometer is subjected during variable-head permeability tests are even more complex because they are non-monotonic. Conventional linear analyses of this type of test under inflow conditions – because of the non-linear behaviour of the soil – lead to coefficients of consolidation and permeability between those pertinent to reloading and primary loading. Variable-head tests with outward flow lead to $k_h(c_h)$ values appropriate for swelling–reloading.

5.4.2 Self-boring instruments

The problems of soil disturbance and smear caused by the installation of piezometers may be overcome to a great extent by the use of self-boring equipment. With the self-boring permeameter (Figure 22) (Jézéquel and Mieussens, 1975), it is possible to determine k_h and c_h using constant-head permeability tests. An example of the measurement of k_h by this device is shown in Figure 23 for Porto Tolle silty clay (Jamiolkowski, Lancellotta and Tordella, 1980). The large scatter of the measured values is attributed to the presence of a highly developed macrofabric (Battaglio *et al.*, 1981).

The difficulty in interpretation of variable-head tests means that constant-head tests are preferred. According to Tremblay (1983), such tests give the most realistic results. Some of the difficulties of comparing inflow and outflow tests, even at constant head, are described in the previous section 5.4.1. Most self-boring permeameter tests have been with outward flow, and it is advisable to perform the tests under only small values of applied hydraulic head. Even with this precaution, k_h and c_h are determined in a stress range near or slightly below σ'_{vo}. Thus, they are not representative of normally consolidated conditions, and the result is better suited for field problems where $\Delta\sigma_v + \sigma'_{vo} \leqslant \sigma'_p$. On the other hand, with a servo-controlled pump system (Mieussens and Ducasse, 1977), it is possible to conduct inflow tests under a constant head. Tremblay (1983) and Tavenas, Tremblay and Leroueil (1983) report

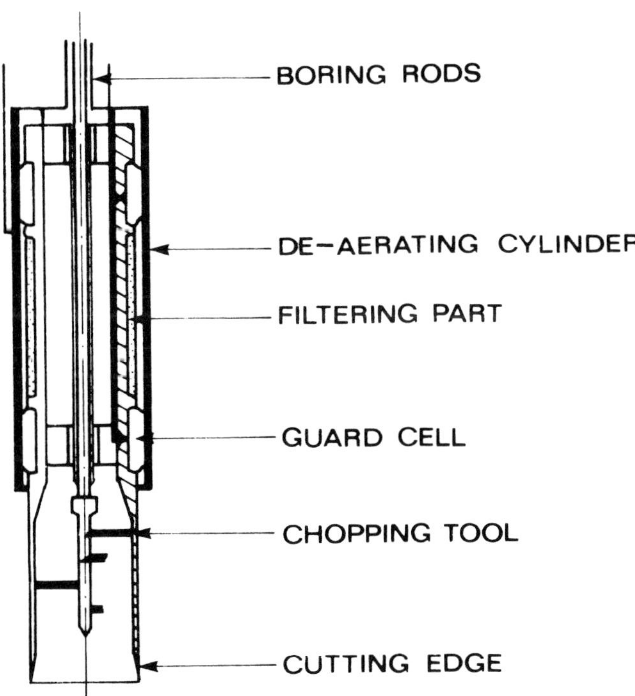

Figure 22 A self-boring permeameter (Jézéquel and Mieussens, 1975)

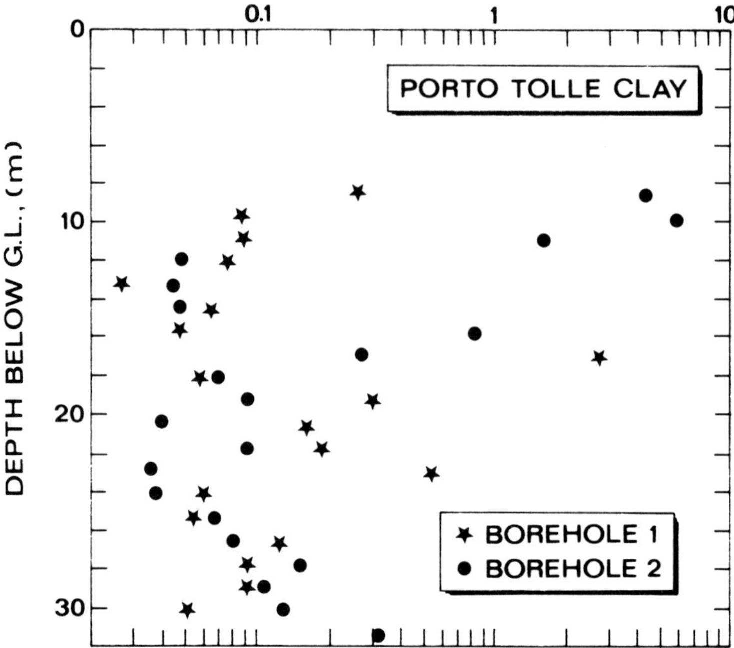

Figure 23 The coefficient of permeability as determined from self-boring permeameter tests (Jamiolkowski, Lancellotta and Tordella, 1980)

on an improved self-boring apparatus which is less susceptible to clogging of the porous element by clay smear during insertion. They also use a very simple system for maintaining a constant head, but the tests are still the outflow type. Thus, the recommendation given above concerning small applied heads is appropriate.

As noted in Section 5.4.1, there are problems in the interpretation of all *in-situ* permeability tests because of changes in effective stress during the test. Sorting out the complex consolidation and swelling processes involved is an important research need.

Another promising example of the application of the self-boring technique for the determination of c_h has been presented by Clarke, Carter and Wroth (1979). Using the Cambridge self-boring pressuremeter (Figure 24), c_h was determined by means of so-called 'holding tests' in which the pressuremeter probe was expanded to a prefixed level of circumferential strain (6–10%), and thereafter the applied cavity pressure was adjusted to keep the radius of the probe constant with time.

The decay of the excess pore pressure, which was induced by the expansion of the probe, was monitored by two pore-pressure transducers located at mid-height of the cavity wall. Thus, an entire dissipation curve could be determined. The interpretation of holding tests is based on the theory of the expanding cylindrical cavity (Vesić, 1972). Details of the assumptions required for the interpretation of this test may be

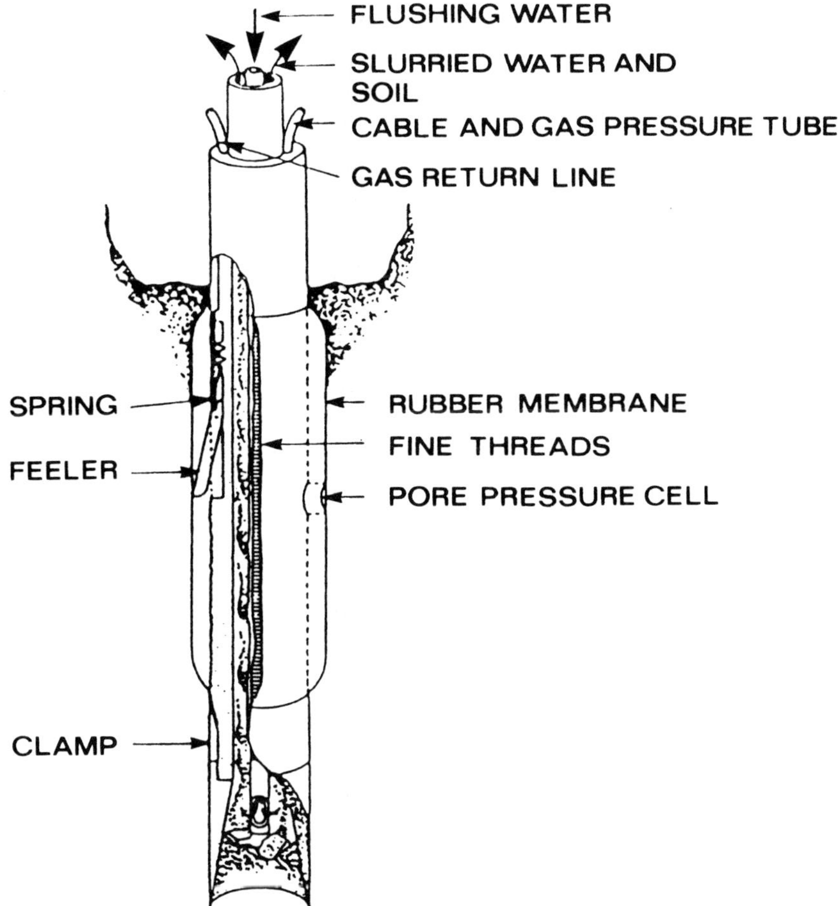

Figure 24 The Cambridge self-boring pressuremeter

found in Carter, Randolph and Wroth (1979) and Clarke, Carter and Wroth (1979). Results of holding tests performed by Clarke (1981) and Benoit (1983) are summarised by Jamiolkowski *et al.* (1985).

The surrounding soil is subject to non-monotonic changes of effective stress during holding tests, and therefore one can expect that, by analogy of what happens during piezocone dissipation tests, the obtained values of c_h are somewhere between those corresponding to conditions of reloading and primary loading.

5.5 Consolidation and permeability characteristics: piezocone dissipation tests

During the last ten years, much attention has been devoted to dissipation tests in clays performed with the piezocone cone penetration test (CPTu) or the pore-pressure probe (PPP) (Figure 25). At certain points during the penetration of the cone, the test is stopped and the decrease of the excess pore pressure with time is monitored (Figure 26). From these readings, an approximate value of the coefficient of consolidation can be obtained. This test has the potential of providing a cost-effective tool for obtaining the variation with depth of the flow and consolidation characteristics at a site.

A simplified approach for the interpretation of piezocone dissipation records was proposed by Torstensson (1975). The soil is assumed to behave as an elastic, perfectly plastic, material subjected to isotropic initial stress. The initial excess pore-pressure distribution is estimated by one-dimensional (spherical or cylindrical) cavity expansion theories (Vesić, 1972), and the consolidation process is analysed by linear-

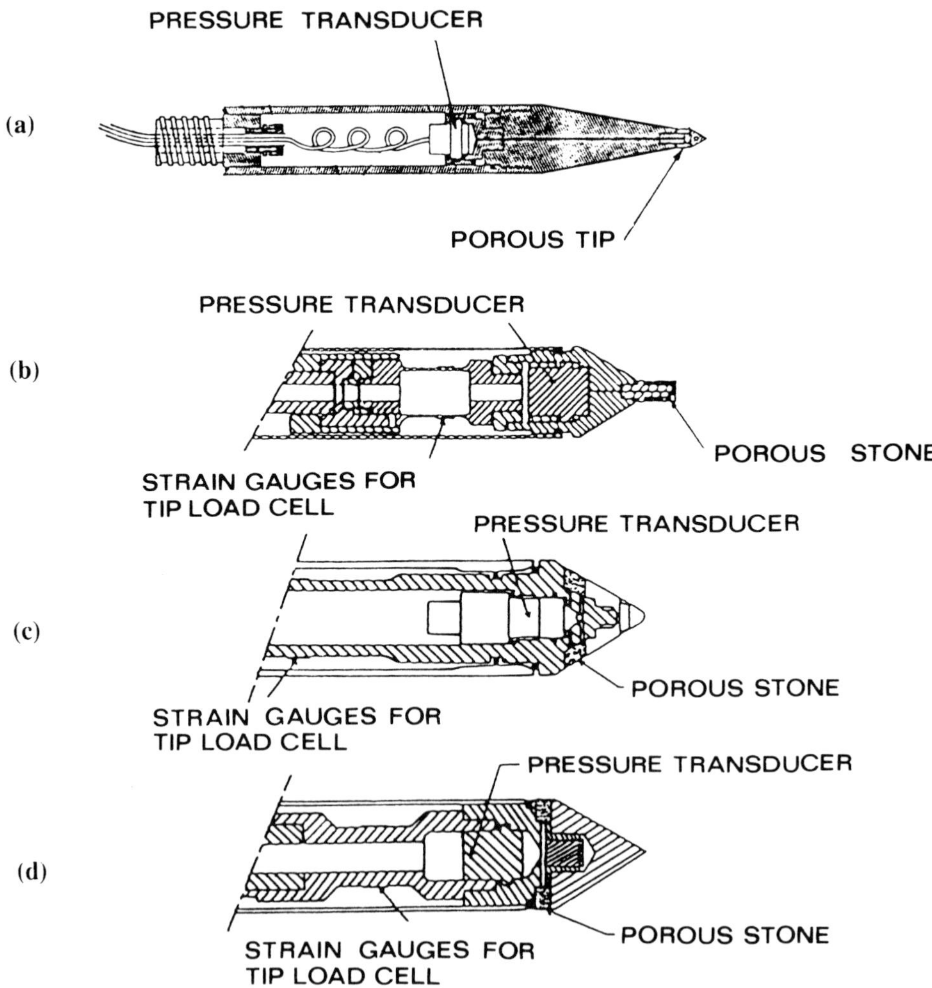

Figure 25 Examples of pore-pressure probes and piezocones

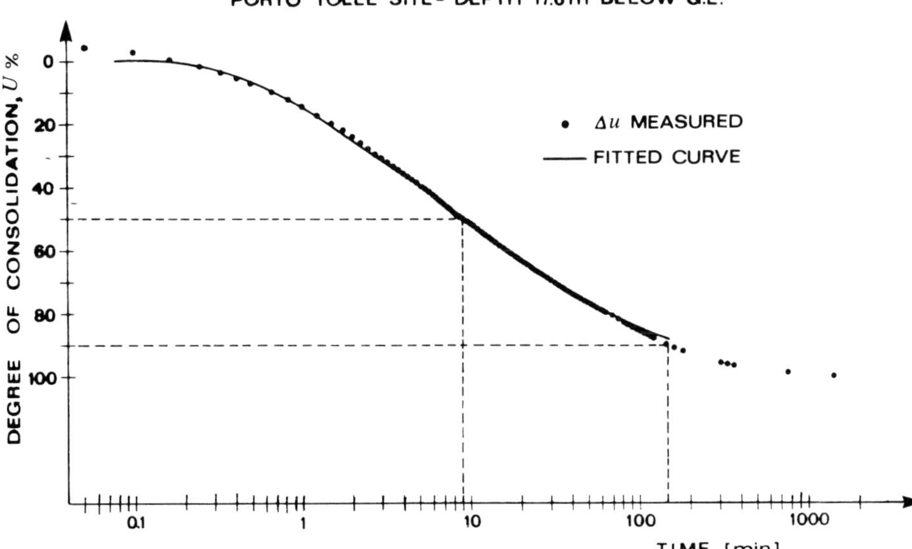

Figure 26 An example of a piezocone dissipation test (Battaglio *et al.*, 1981)

uncoupled one-dimensional theory. The coefficient of consolidation, c, can be obtained from:

$$c = (TR_c^2/t) \tag{14}$$

where T = the time factor for a given degree of consolidation (usually taken as 50%)
t = the measured time to reach the corresponding excess pore-pressure decay
R_c = the equivalent cavity radius

The solution is given in Figures 27a and 27b which show that the consolidation rate is dependent on the shape of the cavity (cylindrical or spherical), its radius, R_c, the pore-pressure coefficient at failure, A_f and the rigidity index, I_R:

$$I_R = G_u/c_u \tag{15}$$

where c_u = the undrained shear strength of the clay
G_u = the undrained shear modulus, usually assumed equal to G_{u50}, which is the secant modulus at half the applied principal stress difference at failure

For natural clay deposits, the rigidity index, I_R, ranges from 50 to 300; it decreases with increasing OCR, and, for the same OCR, it increases with decreasing plasticity index. Self-boring pressuremeter tests yield I_R values generally larger than those resulting from laboratory shear tests (Table 14).

The coefficient of consolidation, c, determined from Equation (14) is c_h if a cylindrical cavity is assumed. If a spherical cavity is assumed, then c is an average value lying somewhere between c_h and c_v. Unfortunately, it is difficult to decide which

Table 14 Rigidity indices of some natural soft-clay deposits (Ghionna *et al.*, 1981)

Site	Plasticity index	Overconsolidation ratio (OCR)	Rigidity index I_R (=G_{u50}/c_u)		Sensitivity from field vane tests
			Laboratory shear tests	Self-boring pressuremeter	
Trieste (Italy)	47±10	1	170	206	3–4
Porto Tolle (Italy)	31±2	1	110	290	2–3
Panigaglia (Italy)	45–65	1–1.2	135	478	3–5
Drammen (plastic clay) (Norway)	25–30	1.2–1.5	203	184	7–8
Onsøy (lower OCR) (Norway)	20–35	1.2–1.7	173	365	5–6
Bandar Abbas (Iran)	33±6	1.5–2	—	133	2–4

G_{u50} and c_u determined from the stress–strain curve as derived from SBP tests; laboratory shear tests: K_o-consolidated, undrained direct simple shear tests

cavity type should be used, particularly when the pore pressure is measured on the cone or just behind it. Cone penetration is a two-dimensional problem, whereas the cavity expansion solutions are one-dimensional. Spherical cavity expansion predicts higher c values than cylindrical (Figure 27) and the ratio between them is about 5.

Because of the uncertainty as to cavity shape, it is also unclear which radius (of the tip or of the porous stone, etc.) should be used when computing c. For example, with Torstensson's (1975) PPP geometry, the two limiting possibilities (i.e. the radius of the porous stone and that of the shaft) result in an uncertainty of about a factor of 4 in the computed c.

Even the cavity expansion theories themselves are subject to criticism (Tavenas, Leroueil and Roy, 1982). For example, the theories assume constant and unique values of E_u and c_u, but it is known that these soil properties are stress-path-dependent. Strain-softening behaviour is neglected, and penetration rate effects are not accounted for by the theories. Because of the very complex nature of the penetration process, pore-pressure dissipation around the CPTu and PPP tips occurs first in remoulded or partly disturbed clay, then in intact clay at some unknown distance from the probe (Tavenas, Leroueil and Roy, 1982). Moreover, during the consolidation process, the excess pore pressure decreases with time, so that soil elements adjacent to the tip are subjected to monotonic loading because of the presence of the rigid boundary. However, those elements away from the tip are first unloaded as the excess pore pressure increases, and then reloaded as the excess pore pressure decreases (Carter, Randolph and Wroth, 1979). Therefore, the consolidation coefficient from this test should not be considered a unique value; further, it is not clear what is the meaning of the c value predicted by a simplified approach to piezocone interpretation.

In spite of all these questions, Robertson *et al.* (1986) showed that the rates of settlement of a preload fill on an organic clayey silt with prefabricated drains could be predicted quite well using c_h values obtained from piezocone dissipation tests.

The interpretation of piezocone dissipation tests for evaluation of consolidation

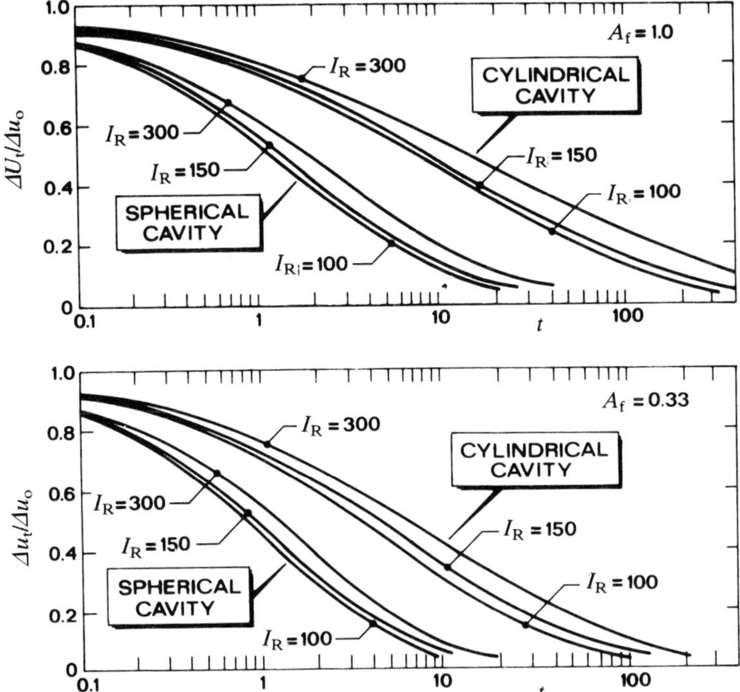

Figure 27 Pore-pressure decay versus time factor, using both cylindrical and spherical expanding cavity theories (Battaglio *et al.*, 1981)

coefficients is still subject to much uncertainty, and therefore use of the obtained c values for design requires judgement and caution. At present, the most comprehensive treatment of this problem has been by Baligh and Levadoux (1980). These authors, using the 'strain-path method' (Baligh, 1985), have been able to take into account the two-dimensional, axisymmetric nature of the penetration process, which appears to lead to a more realistic prediction of the initial pore-pressure isochrone. From both the theoretical and practical points of view, the most relevant conclusions of this work are as follows.

(1) The effect of coupling between total stresses and pore pressure is small, except at the early state of consolidation (pore-pressure decay less than 20%) and near a cone with an apex of 18°. Thus, the simpler uncoupled solution provides reasonably accurate predictions of the dissipation process.
(2) A two-dimensional analysis of consolidation around the cone shows that the dissipation rate is mainly controlled by c_h, and that even a tenfold change of c_v has a negligible influence on the shape of the isochrones. Hence, the test really yields c_h values.
(3) The cone's penetration produces undrained shearing of the soil with a pore-pressure increase and a reduction of the effective stresses. When these excess

pore pressures start to dissipate, the soil surrounding the cone is subject to an increase of effective stresses under conditions of reloading, and only after some dissipation has taken place do the effective stresses equal those existing before tip penetration.

Only from this point onward does the consolidation proceed along the virgin compression curve. As a consequence, Baligh and Levadoux (1980) postulated that c_h obtained from the early stages of dissipation (less than 50% consolidation) is relevant for reloading conditions which therefore reflect the behaviour of overconsolidated soils.

(4) For the locations of the porous stone investigated (Figure 28), Baligh and Levadoux (1980) showed that in normally consolidated and slightly overconsolidated Boston Blue Clay:
   (a) for an 18° cone, the dissipation rate is influenced by the position of the filter;
   (b) for a 60° cone, the filter located on the cone face or at its base gives substantially equal dissipation rates;
   (c) for a filter located far behind the cone, the dissipation rate is no longer influenced by the tip geometry, but the rate is appreciably slower.

Based on these findings, Baligh and Levadoux (1980) recommend the following tentative procedure for the evaluation of c_h from the CPTu or PPP dissipation tests.

(1) The normalised excess pore pressure, $\Delta u_{(t)}/\Delta u_{\infty}$, should be plotted against the time factor, T, on a log scale (Figure 29). In this case, Δu_{∞} = initial excess pore pressure and $\Delta u_{(t)}$ = excess pore pressure at a given time.

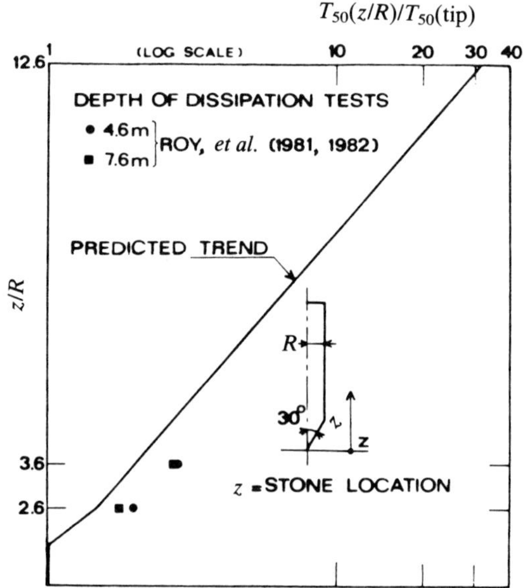

Figure 28 Predicted and measured dissipation rates for different locations of the porous stone (Baligh and Levadoux, 1980 (predicted rates) and Roy *et al.*, 1981, 1982 (measured rates))

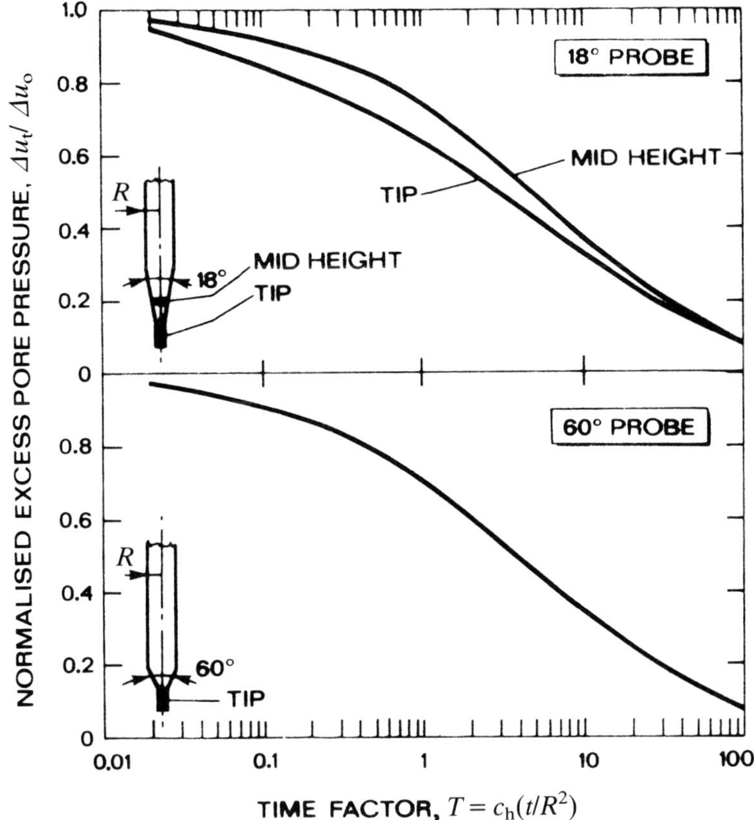

Figure 29 Pore-pressure decay/time factor, using the strain-path method (Baligh and Levadoux, 1980)

(2) The measured dissipation curves should be of similar shape with respect to the corresponding theoretical ones in Figure 27. In fact, according to linear consolidation theory, a horizontal translation of the dissipation curves reflects only changes in c_h, whereas their overall shape depends strongly on the initial isochrone. Therefore, if the condition of similar-shaped curves is not fulfilled, the initial isochrone in the field differs from the theoretical one assumed by Baligh and Levadoux (1980).

(3) When the observed dissipation curve is of similar slope to the theoretical one, c_h may be computed from:

$$c_h = (R_r^2 T)/t \tag{16}$$

where c_h = coefficient of consolidation for horizontal pore-water flow from the CPTu or PPP dissipation test

R_r = radius of the pushing rods

T = time factor for the appropriate tip geometry and porous stone location
t = time elapsed to reach a given degree of consolidation, U

Table 15 and Figure 29 provide values of T for different probe types and various degrees of consolidation.

(4) The value of c_h determined for $\Delta u_{(t)}/\Delta u_o = 0.5$ (i.e. a degree of consolidation of 50%) may be used in problems involving horizontal flow in the overconsolidated range. In order to obtain the coefficient of consolidation in the normally consolidated range, the following rule is suggested:

$$c_{h(NC)} = \left(\frac{RR}{CR}\right) c_{h(OC)} \tag{17}$$

where RR = recompression ratio
CR = virgin compression ratio
NC = normally consolidated range
OC = overconsolidated range

For some typical values of the ratio CR/RR, see Table 16.

(5) Approximate estimates of the horizontal coefficient of permeability, k_h, can be obtained from the expression:

$$k_h = \frac{\gamma_w}{2.3\sigma'_{vo}}(RR)c_{h(OC)} \tag{18}$$

where γ_w is the unit weight of water.

(6) The coefficient of consolidation for vertical flow may be determined approximately from Equation (19).

$$c_v = (k_v/k_h)c_h \tag{19}$$

A rough estimate of the *in-situ* anisotropy in permeability of clays (expressed by the ratio k_h/k_v) can be made using the data given in Table 17.

Table 15 Time factors (Baligh and Levadoux, 1980)

Degree of consolidation ($\Delta u_{(t)}/\Delta u_o$)	*18° cone with filter at*		*60° cone with filter at tip**
	Tip	*Mid-cone*	
0.6	1.4	2.6	1.9
0.5	3.0	4.7	3.65
0.4	6.0	8.2	6.5
0.2	30	34	27

*Applicable also to the filters located at mid-cone and at cone base.

Table 16 Typical values of recompression ratio and virgin compression ratio for soft clays

Site	*Plasticity index*	*RR*	*CR/RR*	*No. of tests*
Porto Tolle (Italy)	31 ± 12	0.0031	6.5	95
Panigaglia (Italy)	45–65	0.025	8.0	33
Trieste (Italy)	47 ± 10	0.032	6.6	43

Table 17 Range of possible values of the ratio of coefficients of permeability for horizontal/vertical flow for soft clays

Nature of clay	*Permeability ratio (k_h/k_v)*
No macrofabric, or only slightly developed macrofabric, essentially homogeneous deposits	1–1.5
From fairly well to well-developed macrofabric, e.g. sedimentary clays with discontinuous lenses and layers of more permeable material	2–4
Varved clays	2–3
Other deposits containing embedded and more or less continuous permeable layers	3–15

The preceding analysis of CPTu dissipation tests by Baligh and Levadoux (1980) does not include any possible influence of smear, which may lead, if not properly considered, to an underestimate of c_h. Acar, Tumay and Chan (1982) and Fioravante (1983) studied this problem by numerical parametric analyses. Their results suggest that, because of all the problems associated with the interpretation of CPTu dissipation records, the influence of smear may be relevant only in sensitive soils with very anisotropic permeability. In that extreme case, neglecting the presence of a smeared zone of radius $1.5R_r$ and assuming a five-fold permeability reduction, results in an underestimate of c_h by a factor between 2 and 3. In low-sensitivity, cohesive deposits, this factor is much lower, ranging from 1.3 to 1.5.

As well as the theoretical aspects of the analysis of CPTu dissipation, interpreting and using the records successfully in practice requires a rigid and extremely well de-aired pore-pressure measuring system (Gillespie and Campanella, 1981; Jamiolkowski *et al.*, 1985).

5.6 Consolidation and permeability characteristics: back-analysis from field measurements

In view of all the uncertainties involved in the determination of design values of c_h and k_h from laboratory and *in-situ* tests, back-analysis of the excess pore pressures and displacements measured in full-scale structures or under trial embankments may be considered, at least in principle, as the most reliable way to assess flow and consolidation properties. In fact, for large vertical-drain projects on deposits with significant macrofabric, it was recommended earlier (Table 13) that trial embankments should be built to aid the prediction of rates of settlement and pore-pressure dissipation. Pore pressure and settlement observations provide an excellent opportunity to estimate the coefficient of consolidation of the foundation soil and to check the performance of preselected drains. Also, as the durability and mechanical properties of new prefabricated drains may be questioned, observation of their field performance is the best way to show that a selected drain is appropriate for the specific load and site conditions. However, back-analysis always involves a number of simplifying assumptions (e.g. boundary and/or drainage conditions, soil behavioural model, auxiliary parameters which have to be introduced into the analysis, etc.) which may compromise the reliability of the computed values of c or k. Their influence on the obtained results should always be evaluated. In this respect, it is important to remember the following.

(1) Soil parameters obtained from field measurements can be compared with the results of *in-situ* tests only when they reflect the performance at a single point in the soil mass (i.e. pore-pressure measurements or locally measured strains), whereas those obtained from the analysis of average conditions (i.e. settlement records) are representative of overall or average behaviour.
(2) The methodology by which the soil parameters are determined depends on the way they are going to be used, i.e. the parameters and the methods of analysis are strictly interrelated (Lambe, 1973).

Estimates of c_h from back-analyses may be based on either pore-pressure or settlement measurements.

5.7 Back-analysis by pore-pressure observations

The evaluation based on pore-pressure observations involves the following steps.

(1) Determine from field observations or compute from theory the value of the initial excess pore pressure Δu_o (Parry and Wroth, 1977; Hoëg, Andersland and Rolfsen, 1969; Leroueil *et al.*, 1978; Jamiolkowski and Lancellotta, 1984).
(2) Compute the excess pore pressure, $\Delta u_{(t)}$, from piezometer readings, and determine ratios of $\Delta u / \Delta u_o$ for different times, t. Use them with the appropriate theory to compute the non-dimensional time factor, T_h, for horizontal or radial consolidation.

(3) Evaluate c_v (or c_h) using either the total-time method or the incremental-time method (Bromwell and Lambe, 1968). In the first case, the expression is:

$$c_h = (T_h d_e^2)/t \tag{20}$$

and the second case:

$$c_h = \frac{T_{h2} - T_{h1}}{t_2 - t_1} d_e^2 \tag{21}$$

where the terms are as defined in Chapter 3, Section 3.2.

According to Bromwell and Lambe (1968), the use of the incremental-time method tends to minimise the effects of any errors in the assumption of Δu_o.

An alternative to the above approach for the evaluation of c_h is fitting the curve of the observed decay of excess pore pressure with time using the theoretical relationships developed by Barron (1948) for radial drainage and equal strain conditions. In this case, an average value of c_h is appropriate for the observed consolidation process.

The equations are also given in Jamiolkowski, Lancellotta and Wolski (1983a). However, there are three difficulties with the evaluation of c_h using pore-pressure observations; these are considered below.

5.7.1 Assessment of initial excess pore pressure, Δu_o

The value Δu_o, being measured at the end of loading, reflects the rate of trial embankment construction and does not necessarily represent the true initial excess pore pressure.

Factors like partial consolidation during construction, installation procedure (e.g. vibratory equipment (Lee, 1983)) and small or not fully recognised OCR values of the foundation soil, can also affect the measured Δu_o value.

Therefore, Foott and Ladd (1973) and Parry and Wroth (1977) recommend that the initial pore-pressure increase should be related to the corresponding total stress increment on the basis of the theoretical considerations. Alternatively, semi-empirical procedures developed by Leroueil *et al.* (1978) could be used; the resulting Δu_o values are then compared with those measured. If the differences between them are significant, it is advisable to base the evaluation of c_h on both values of Δu_o.

When evaluating Δu_o as a function of the total stress increment, it has to be kept in mind that its magnitude also depends on factors like confined plastic flow and undrained creep, which are difficult to evaluate. Proper consideration of these factors in embankment analyses is an important research need.

5.7.2 Piezometer and drain locations

The use of pore-pressure readings is very sensitive to the assumed locations of the piezometer tip with respect to drain positions. Unfortunately, exact positions are not known because neither the drains nor the piezometers are likely to have been installed precisely vertical. This factor may be particularly important in the case of long prefabricated drains installed with a flexible mandrel and with piezometers at large depths.

Very little is known about these deviations from the vertical, although they have been mentioned by Hansbo (1960), Choa *et al.* (1981), Nicholson (1982) and Magnan and Deroy (1980). On the basis of the available information, deviations from the intended position can be estimated as follows.

(1) Prefabricated drains of small dimensions: within a radius of about 0.025 times installed length.
(2) Piezometers installed in boreholes: within a radius of about 0.01 times installed length, but this value may be larger in the case of piezometers which have been pushed in.

In addition to the above deviations, errors in placement at the ground level have to be considered.

5.7.3 The stress history of the deposit

In cases where $\sigma'_{vo} < \sigma'_p < (\sigma'_{vo} + \Delta\sigma_v)$ the consolidation coefficients c_v and c_h change significantly during the consolidation process, and this makes the evaluation of c_h from pore-pressure data very complex. In fact, under these circumstances, the first part of the settlement corresponds to recompression, and this implies a very rapid initial consolidation because the c_h values are large in the overconsolidated range. Also in this range, $\Delta u_\infty \simeq \Delta\sigma_{oct}$, (the change in octahedral stress). When the applied vertical stress exceeds σ'_p the consolidation rate decreases because of the reduced value of c_h in the normally consolidated range.

In addition, the incremental Δu_∞ values tend to be appreciably higher than $\Delta\sigma_{oct}$, because of the development of plastic zones and the related contribution of octahedral shear stress to the generated excess pore pressures. This non-linear soil response makes the interpretation of piezometer readings more complex. It may be preferable to refer the analysis to that instant at which the applied vertical stress exceeds σ'_p and to use the incremental-time approach. For further details see Noiray (1982).

An example of the difficulties involved in such analyses, mainly in evaluating the initial excess pore-pressure response, Δu_o, was given by Jamiolkowski and Lancellotta (1984). They examined the behaviour of a trial embankment on four different types of vertical drains, and they concluded as follows.

(1) The elasto-plastic approach leads to unrealistic results. It predicts values of Δu_o lower than, or equal to, that measured at the end of the loading ramp, Δu_{max}. This is particularly unsatisfactory in the presence of vertical drains.
(2) The empirical and semi-empirical approaches yield more realistic comparisons between Δu_o and Δu_{max}, but they still lead to a degree of consolidation at the end of the loading ramp which appears too low if it is compared to that deduced from observed consolidation settlements.

(3) The evaluation of Δu_o using the modified Cam-Clay model leads to $\Delta u_{max}/\Delta u_o$ ratios which are in agreement with the values of the consolidation settlement.

The ratios of $\Delta u_o/\Delta \sigma_v$ which are consistently greater than unity may be attributed, at least partially, to the occurrence of constrained plastic flow and strain-softening phenomena.

5.8 Back-analysis from settlement observations

The other way to estimate c_h by back-analysis is to use field settlement observations. In this case, the ratio $s_{c(t)}/s_{cf}$ is used

where $s_{c(t)}$ = measured consolidation settlement of the same layer at any intermediate time
s_{cf} = computed final consolidation settlement of the layer in which consolidation is accelerated by vertical drains

The evaluation steps are analogous to the ones taken when c_h is evaluated from pore-pressure observations, as outlined above. As with that approach, the one based on settlement observations also suffers from a number of uncertainties, as follows.

(1) Obtaining a reliable estimate of s_{cf} on the basis of laboratory tests is not easy. The best accuracy that can be expected is around $\pm 20\%$ (Holtz and Broms, 1972). This difficulty may partly be overcome by extrapolating the observed field settlement curves to obtain a probable value of s_{cf}, as shown by Asaoka (1978) and Magnan and Deroy (1980).
(2) In order to assess the value of $s_{c(t)}$ at a given moment, it is necessary to deduct all settlements resulting from the compression of the layers above and below the one in which consolidation is accelerated by vertical drains. This requires reliable settlement measurements at different depths (Hansbo, 1960), and thus a degree of uncertainty in the assessment of $s_{c(t)}$ is introduced.
(3) It is difficult to determine which part of the measured settlement results from the consolidation alone. This means that the vertical displacement from the immediate settlement, s_i, (including both the 'elastic' and 'confined plastic' flow components) should be subtracted from the value of $s_{(t)}$ measured at a given time, t.

Yet as noted by d'Appolonia, Poulos and Ladd (1971), Tavenas *et al.* (1974), and Tavenas and Leroueil (1980), good estimates of s_i are difficult to make because of problems in reliably estimating E_u and ν, and because of partial drainage occurring during construction. The actual behaviour of embankments is complex and apparently differs from what is usually assumed, i.e. undrained behaviour during construction and then drained response thereafter in the long term (Tavenas and Leroueil, 1980). Specifically, when appreciable consolidation settlement is expected to have occurred during construction, it is not correct to assume that the end of construction settlement corresponds entirely to s_i. This is particularly true if the soil is overconsolidated or even slightly overconsolidated, e.g. as shown by Leroueil *et al.* (1978) and Tavenas and Leroueil (1980). If granular layers are included in the soil profile, their contribution to s_i also has to be estimated (Hegg *et al.*, 1983).

The effect on vertical settlement of large horizontal movements which occur because of overstressed zones near the edges of the embankment was considered by Holtz and Lindskog (1972), Noiray (1982), and Hegg *et al.* (1983). Their work does not solve the problem completely nor do procedures such as SHANSEP (Ladd and Foott, 1974) or finite element methods help (Tavenas *et al.*, 1974; Tavenas and Leroueil, 1980). These latter authors recommend, because of these difficulties, a largely empirical approach, which is simpler than other existing methods and because it is based on published case records, is also quite reliable. Unfortunately, the cases reviewed by Tavenas and Leroueil (1980) do not include any with vertical drains and, thus, the effect of this drainage on their procedure is unknown.

(4) With overconsolidated soil (i.e. if $\sigma'_p > \sigma'_{vo}$) the estimation of $c_h = f(s_{c(t)}/s_{cf})$ is even more complex. In this case, by analogy with what was said above with regard to the evaluation of c_h from pore-pressure data, it is advisable to subtract the consolidation settlement from recompression (up to the *in-situ* value of σ'_p) from both $s_{c(t)}$ and s_{cf}. Thereafter, the incremental method (Bromwell and Lambe, 1968) can be applied, preferably with respect to primary consolidation data only.

(5) When evaluating $c_h = f(s_{c(t)}/s_{cf})$, the implicit assumption is made that, under conditions of partial consolidation, the vertical settlement from creep is neglected. (This is true also with procedures for estimating s_i.) In cases where this is not true, one tends to overestimate c_h.

5.9 Systematic analysis of field data

In spite of all the difficulties with back-analysis using field pore-pressure measurements and settlement observations just described, a few procedures have been developed for a systematic analysis of field data which help to overcome some of these problems. Johnson (1970a, 1970b) and Magnan (1983) reviewed these methods in some detail, and both give charts for determining c_v (c_h) from pore-pressure observations. Methods based on field settlement observations which have been applied successfully to vertical-drain situations include those by Long and Carey (1978) and Asaoka (1978). Both use simple graphical procedures and calculations to determine c_h.

When evaluating c_h from settlement data, the procedure suggested by Asaoka (1978) is recommended (Figure 30). This method is based on Barron's (1948) solution for pure radial drainage. The relevant steps can be summarised as follows.

(1) From the time–settlement curve, select a series of settlement values $s_1, s_2 \ldots s_n$,

(a) SETTLEMENT VS. TIME

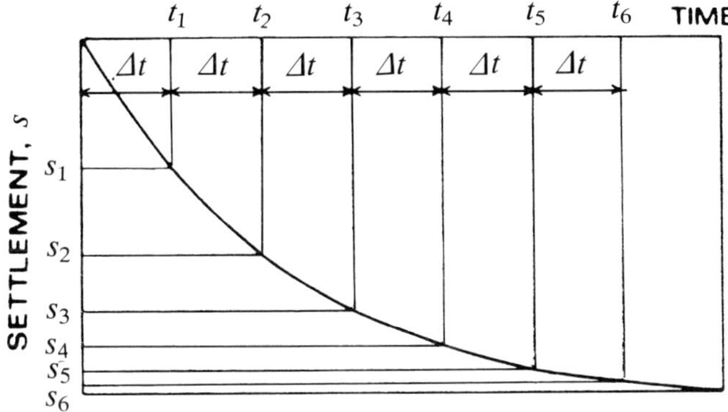

(b) ASAOKA'S (1978) CONSTRUCTION

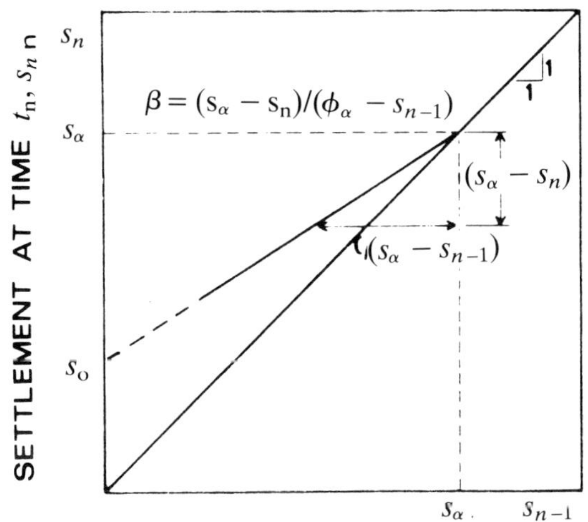

SETTLEMENT AT TIME t_{n-1}, s_{n-1}n−1

Figure 30 An example of Asaoka's construction (Asaoka, 1978 (from Orleach, 1983))

such that s_n is the settlement at time t_n and that the time interval $\Delta t = (t_n - t_{n-1})$ is constant (Figure 30a).

(2) From this series, plot the points (s_{n-1}, s_n), as shown in Figure 30b.
All points should lie on a straight line, such that:

$$s_n = s_o + \beta s_{n-1}$$

where s_o and β are two constants which depend on the selected time interval Δt.

(3) The constant, β, represents the slope of the constructed straight line, and it can be related to the coefficient of consolidation by:

$$c_h = -\frac{d_e^2 \, F(n) \log_e \beta}{8 \Delta t} \qquad (22)$$

The value of c_h is theoretically independent of the chosen time interval and the time of origin.

The same approach can be used for consolidation with vertical drainage only. The relation between the slope β and c_v is (Magnan and Deroy, 1980; Magnan, 1983):

$$c_v = -\frac{4 H_d^2 \log_e \beta}{\pi^2 \Delta t}$$

This relationship is only valid for values of the time factor for vertical flow $T_v > 0.1$.

It is stressed that Asaoka's (1978) procedure yields a straight line in Figure 30b only if the soil behaviour fulfils the assumptions of the Terzaghi theory. Actual data (Figure 31) may have an initial or final upward curvature in the following circumstances.

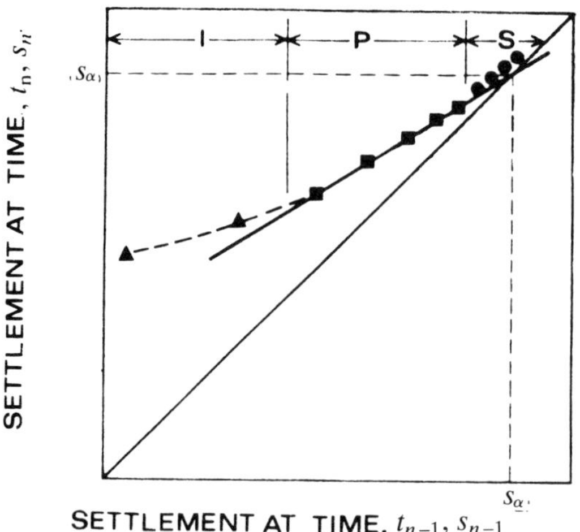

ZONE	SYMBOL	STATE
I	▲	INITIAL STAGE OF CONSOLIDATION (c_h OR c_v DECREASING)
P	■	PRIMARY CONSOLIDATION WITH CONSTANT c_h OR c_v)
S	●	SECONDARY COMPRESSION

Figure 31 A possible deviation from Asaoka's construction (Orleach, 1983)

(1) The condition of one-dimensional pore-water flow is not fulfilled.
(2) c_h or c_v changes during the consolidation process.
(3) Secondary compression occurs.

This approach was applied by Orleach (1983) to experimental data obtained from a prefabricated-drain project in Mobile, Alabama. The results are shown in Table 18, and the comparison with those from pore-pressure data confirm the general trends found in the literature, as follows.

(1) The field c_h is considerably higher than c_h from conventional laboratory tests.
(2) The c_h back-calculated from pore-pressure records is lower than those computed from settlement records (Ladd, Rixner and Gifford, 1972; Holtz and Broms, 1972; Adachi and Todo, 1979; Choa *et al.*, 1981). This can be explained partly by theory (i.e. compare coupled and uncoupled consolidation theory), but also by field instrumentation performance (Orleach, 1983).

Other examples in which the Asaoka–Orleach procedure was used with apparent success have been reported by Magnan and Deroy (1980), Magnan, Pilot and Queyroi (1983) and Jamiolkowski *et al.* (1985) (see also Chapter 3, Section 3.3).

Finally, when dealing with data from either excess pore-pressure or settlement records from cases where vertical drains have been installed, it is necessary to evaluate the smear effect in order to assess a reliable value of c_h. A procedure for this is given in Section 5.15.

5.10 Stress history

The stress history of a soil deposit plays an important part in the designer's decision whether to attempt improvement of that deposit by preloading and vertical drains. If the deposit has been subjected to mechanical overconsolidation, desiccation, and/or ageing so that $(\sigma'_{vo} + \Delta\sigma_v) \leqslant \sigma'_p$, the use of vertical drains is not necessary, and may even have a detrimental effect on the future performance of the proposed structure.

Table 18 Values of the coefficient of consolidation for horizontal flow obtained from a back-analysis of pore-pressure and settlement data from Mobile, Alabama (Noiray, 1982; Orleach, 1983; also, Asaoka, 1978)

Approach	*From*	*Coefficient of consolidation (horizontal flow), c_h (m²/year)*
Total elapsed time	Pore-pressure data	1.3 ± 0.6
Incremental time	Pore-pressure data	1.3 ± 0.3
Asaoka (1978)	Pore-pressure data	1.4 ± 0.3
Total elapsed time	Settlement data	4.7 ± 1.0
Incremental time	Settlement data	3.4 ± 0.3
Asaoka (1978)	Settlement data	4.4 ± 0.5

Note: Laboratory $c_v = 0.7 \pm 0.3\ m^2/year$.

Generally, in these circumstances, the soil possesses a sufficiently high coefficient of consolidation to ensure that consolidation settlements (which also may be expected to be of limited magnitude) occur in a relatively short time without drains. This point has been discussed in some detail by Bjerrum (1972).

If the clay stratum to be improved is overconsolidated, but the foundation stresses to be induced by the structure will exceed σ_p', the effectiveness of vertical drains depends upon the factor η, defined by Bjerrum (1972) as:

$$\eta = \frac{\log \sigma'_{vf} - \log \sigma'_{p}}{\log \sigma'_{vf} - \log \sigma'_{vo}} \tag{23}$$

where σ_p' = preconsolidation stress

$\sigma'_{vf} = \sigma'_{vo} + \Delta\sigma_v$ = final effective consolidation stress

σ'_{vo} = effective overburden stress

According to Bjerrum (1972), the drain installation is justified when η is higher than 0.6, although drains may still be needed in cases where η is as low as 0.4.

The critical factor in the stress history of a soil is the preconsolidation stress σ_p', which may also be expressed in terms of the OCR. It is most commonly determined in the laboratory by some type of oedometer test on high-quality (undisturbed) samples. It may also be possible to determine σ_p' in the field by using a PPP or CPTu test. Recent developments in both these areas have been discussed by Jamiolkowski *et al.* (1985), and thus only a brief summary is given here.

'Preconsolidation pressure' may be a somewhat misleading term in the physical sense, because there are several mechanisms other than earlier loading which may be the cause of this threshold stress (Brummund, Jonas and Ladd, 1976; Jamiolkowski *et al.*, 1985). Yet it remains the accepted term to denote the change in slope in the oedometer curve. As noted by Jamiolkowski *et al.* (1985), it really represents a yield stress which separates small-strain 'elastic' behaviour and rapid consolidation from large strains accompanied by plastic (irrecoverable) deformation occurring relatively slowly during one-dimensional compression.

Procedures for the determination of σ_p' by the oedometer test are described by Ladd (1971), Holtz and Kovacs (1981), Padfield and Sharrock (1983) and Weltman and Head (1983) among others. The incrementally loaded oedometer test is probably still the most common worldwide test, although the constant rate of strain (CRS) and controlled gradient (CG) tests have their proponents. The CRS test is now the standard oedometer test in both Norway and Sweden (Larsson and Sällfors, 1986; Sandbaeken, Berre and Lacasse, 1986). The CRS and CG tests are continuously loaded, so a continuous curve of e or ε_v plotted against stress is obtained, and thus the change in slope at σ_p' is easily seen. Variables influencing the determination of σ_p' and some of the controversy concerning which compression curve (e.g. end of primary, 24 h consolidation reading, etc.) gives the most reliable results are discussed at some length by Jamiolkowski *et al.* (1985), and in the discussions to Session 2 (Volume V) at the Eleventh International Soil Mechanics and Foundation Engineering Conference in San Francisco in 1985.

At present, there is almost a complete lack of experience concerning the assessment of stress history (OCR or σ'_p) from *in-situ* tests. On the basis of experience with the PPP and CPTu in Boston Blue Clay, Baligh and Vivatrat (1979) and Baligh, Vivatrat and Ladd (1980) suggested that the ratio of the pore pressure measured during penetration to cone resistances, q_c, may reflect the stress history of cohesive soil deposits. The reasoning behind this idea may be found in Baligh, Vivatrat and Ladd (1980).

Using the approach of critical state soil mechanics, Wroth (1984) presented a comprehensive and critical review of the same concept. His analysis may be summarised as follows.

(1) Any pore-pressure coefficient derived from CPTu which can be correlated to OCR should be analogous to that formulated by Henkel (1960):

$$a_f = \frac{\Delta u - \Delta\sigma_{oct}}{\Delta\tau_{oct}} \tag{24}$$

where a_f = Henkel's pore-pressure coefficient at failure
Δu = change in pore pressure
$\Delta\sigma_{oct}$ = change in total octahedral stress
$\Delta\tau_{oct}$ = change in octahedral shear stress

(2) As the actual $\Delta\sigma_{oct}$ and $\Delta\tau_{oct}$ around the CPTu tip are not known, a dimensionless parameter, B_q, proposed by Senneset, Janbu and Svanø (1982) among others, is required, which is a ratio of the measured excess pore pressure to the induced shear stress, or:

$$B_q = \frac{u_{max} - u_o}{q_t - \sigma_{vo}} \tag{25}$$

where u_{max} = penetration pore pressure
u_o = hydrostatic pore pressure
q_t = total cone resistance corrected for the unequal end-area effect
σ_{vo} = total overburden stress

Thus, as stated by Wroth (1984), B_q (probably better reflects the *in-situ* OCR than does u_{max}/q_c, $(u_{max} - u_o)/q_c$, or some of the other ratios proposed.

Experimental evidence of the response of the CPTu to the stress history of clay deposits may be found in Jamiolkowski *et al.* (1985). In addition, more recent piezocone data have been correlated by Coutts (1986) with the OCR for one onshore and five offshore sites in the UK. These results are shown in Figure 32.

As mentioned in the earlier discussion of the CPTu for determining c_h (Section 5.5),

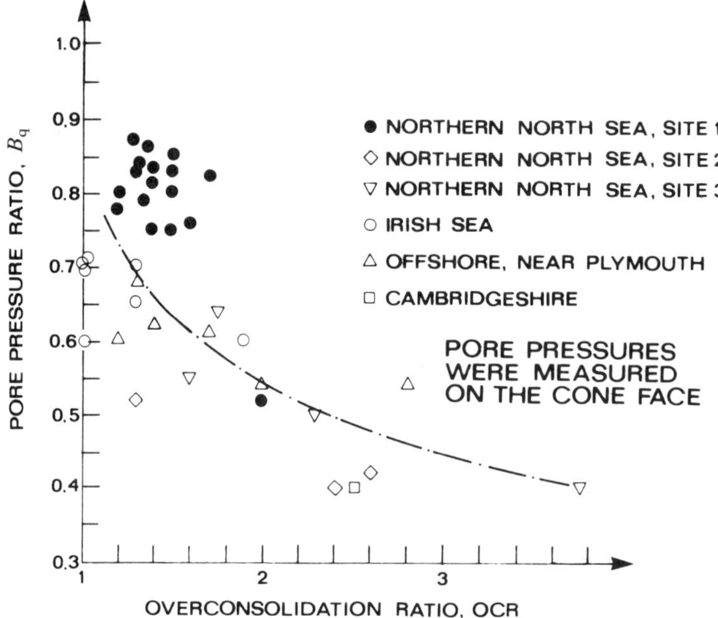

Figure 32 Variation of pore-pressure ratio/overconsolidation ratio for six sites (after Coutts, 1986)

theoretical analyses using undisturbed soil properties such as a_f are questionable, because the clay becomes partially remoulded around the tip. The induced excess pore pressure also depends on a number of other factors such as sensitivity and strain rate.

As in all applications of the CPTu, the question arises as to the optimum position of the filter stone to best reflect the stress history. Arguments for various locations are given by Jamiolkowski *et al.* (1985). Its location at or near the cone base is well suited to reflect the changes in OCR (Wroth, 1984) and this location seems to be an acceptable practical compromise, but there is still considerable research effort and debate on the filter position and the meaning of the recorded pore pressure.

Jamiolkowski *et al.* (1985) reviewed the potential of the CPTu to assess the stress history of cohesive deposits. Their review, which showed just how complex the subject is, led to the following preliminary conclusions.

(1) Based on theory and on soil behaviour observed during laboratory tests, B_q can be expected to reflect OCR changes within a given soil deposit. Most of the examples of $B_q = f$ (OCR) shown in Jamiolkowski *et al.* (1985) tend to validate this statement, but one or two do not.
(2) However, because of differences in sensitivity and in the σ'_p mechanisms among the different soil deposits, a unique relationship between B_q and changes in OCR probably should not be expected.
(3) The problem is made more difficult because soft-soil sites where vertical drains are proposed may be nearly normally consolidated.
(4) In cohesive deposits with a highly developed macrofabric, the use of B_q plotted

with depth as an indicator of OCR seems to be problematic. Further research on this point is needed.

(5) Pore-pressure ratio, B_q, is simultaneously an index of stress history and soil type, and therefore it probably reflects not only OCR changes but also local soil heterogeneity.

(6) Changes in OCR within a given uniform cohesive deposit should be reflected by B_q, but further validation and improved reliability of the $B_q = f\,(\mathrm{OCR})$ correlations, will require:
   (a) a much greater insight into the incremental total stress field around the CPTu tip;
   (b) standardisation of the device and test procedures;
   (c) accumulation of reliable experimental CPTu data from a number of geotechnically well-documented deposits.

5.11 Macrofabric and drainage boundaries

As mentioned in Section 5.3, the effectiveness of vertical drains depends to a large extent on the macrofabric of the deposit, because it is likely the macrofabric determines the drainage-boundary conditions (Rowe, 1968, 1972; Casagrande and Poulos, 1969; McGown *et al.*, 1980; McGown and Hughes, 1981). Therefore, for a technically feasible and economic design, it is essential to gather information on the possible existence of thin layers, lenses or seams of more permeable material which might be embedded in the clay mass. It is equally important to assess if these features are discontinuous or, if continuous, could act as drainage boundaries in the absence of drains. If they are discontinuous, their interconnection by vertical drains generally leads to very efficient drainage and significantly accelerated field consolidation, as noted by Rathmayer (1979) and Jamiolkowski and Lancellotta (1981).

McGown *et al.* (1980), building on Rowe (1972) and their own research at the University of Strathclyde, developed very useful procedures for quantifying macrofabric observations. An example of the application of the system to a vertical-drain project is given by McGown and Hughes (1981) and Dioy, McGown and Mann (1986).

The traditional way of investigating soil macrofabric and drainage boundaries is to carry out soil borings with continuous sampling. Successive piston samples in parallel holes (Hvorslev, 1949) or other special samplers (Kjellman, Cadling and Wager, 1950; Begemann, 1966) can be used. Large-diameter samplers are recommended by Rowe (1968, 1972) and McGown and Hughes (1981). Recently the PPP (Torstensson, 1975; Wissa, Martin and Garlanger, 1975) and the CPTu (Baligh *et al.*, 1981; Tumay, Boggess and Acar, 1981; Campanella and Robertson, 1981; Torstensson, 1982; Roy *et al.*, 1982) have enabled the excess pore pressure generated during cone penetration to be measured. These devices are extremely sensitive, rapid and, therefore, economical tools for investigating soil macrofabric. They are particularly appropriate for the detection of layers, lenses or seams of more permeable material in otherwise impervious cohesive soil deposits. Jamiolkowski *et al.* (1985) summarised developments in the use of the CPTu and PPP devices, including a discussion of the best location of the filter, type of filter, test errors, and overall reliability (see also Meigh (1987) for an overview of the use of piezocones).

Three examples of pore-pressure profiles obtained in Italian clays by Battaglio *et al.* (1981) and Bruzzi and Cestari (1983) are shown in Figure 33. They illustrate the effectiveness of PPPs for the investigation of the macrofabric of cohesive soils. On the Trieste site (Figure 33a) the soft clay layer is almost homogeneous, and no macrofabric features have been detected by the pore pressure. This deposit was

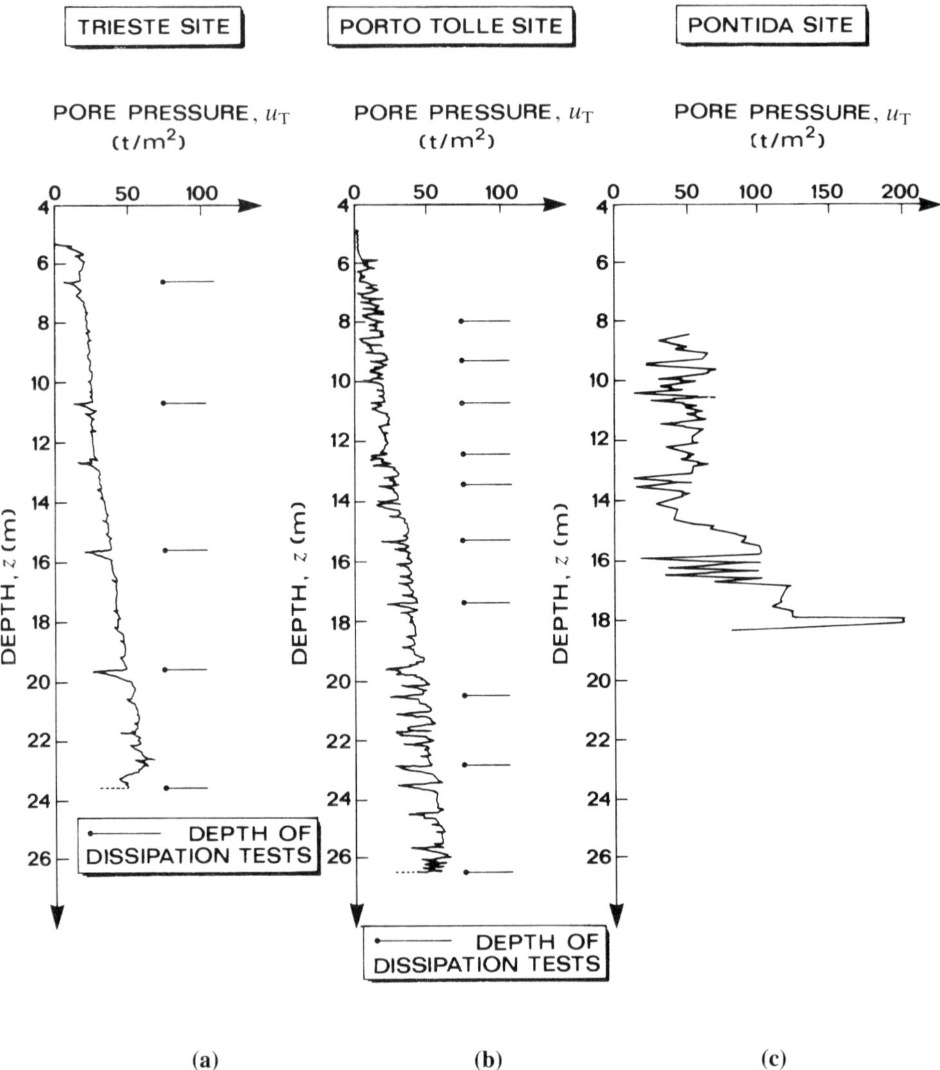

Figure 33 Examples of pore-pressure profiles (Battaglio *et al.*, 1981; Bruzzi and Cestari, 1983)

treated by preloading the vertical sandwick drains, as described by Hansbo, Jamiolkowski and Kok (1981). The field coefficient of consolidation obtained from the back-analysis of piezometer readings was very close to the coefficient of consolidation determined by standard laboratory tests on good-quality undisturbed specimens.

At the Porto Tolle site (Figure 33b), the soft silty clay has a relatively well-developed macrofabric, with lenses and seams of more permeable silty fine sand embedded in the clay mass. This deposit has been subjected to extensive improvement by precompression and the installation of jetted and prefabricated band-shaped drains (Garassino *et al.*, 1979; Hansbo, Jamiolkowski and Kok, 1981; Hegg *et al.*, 1983). In this case, the field coefficient of consolidation determined on the basis of observed pore-pressure and settlement rates was from 1.5 to 8 times higher than the coefficient of consolidation resulting from laboratory tests (Table 19).

Finally, on the Pontida site (Figure 33c), the profile of pore pressures during cone penetration reveals a highly developed macrofabric which consists of almost continuous thin layers of silty sand embedded between layers of silty clay. In this deposit, the consolidation process may be expected to occur rapidly without any vertical drains.

The experience reported by Rowe (1968), Casagrande and Poulos (1969), Choa *et al.* (1981), Jamiolkowski and Lancellotta (1981) and McGown and Hughes (1981) with respect to macrofabric in projects involving vertical drains may be summarised as follows.

(1) The macrofabric of a soil deposit has a large influence on the effectiveness of vertical drains.
(2) It is important, for a rational selection of drain installation equipment, that account is taken of the possible consequences of drain installation on the soil structure (disturbance).

Table 19 A comparison of different determinations of the coefficient of consolidation for horizontal flow of Porto Tolle clay (after Battaglio and Maniscalco, 1983; Jamiolkowski and Lancellotta, 1984)

	c_h *(OC)* *(m²/year)*	c_h *(NC)* *(m²/year)*	*Source*
Piezocone	122 ± 55	19 ± 9	Battaglio and Maniscalco (1983)
Back-analysis of piezometer readings	63–95	—	Jamiolkowski and Lancellotta (1984)
Back-analysis of settlement records	—	15	
Laboratory c_h	9–17		
Field (SB) k + lab m_v	22–32		

(3) It is difficult to quantify the effects of macrofabric in the design of vertical drains or through back-analysis of drain performance, as noted in Sections 5.7 and 5.8.
(4) If there is little or no macrofabric, then as a practical matter c_h may be assumed to be the same as c_v for design. As was shown in Table 17, even with macrofabric, rarely is c_h more than 2–3 times c_v. Therefore, the effect of macrofabric on design, though important, is not extremely so.

5.12 One- and three-dimensional loading

In vertical-drain projects in which the dimensions of the loaded area are small relative to the thickness of the compressible layer, the loading cannot be considered one-dimensional. In this case, significant horizontal deformations can occur at the edge of the embankment, causing vertical displacements which are seen as settlement. Thus, predictions of total settlement using conventional one-dimensional theory would be incorrect. Previous handling of this problem is discussed in Section 5.9.

It is not possible, at present, to determine the contribution of confined creep and plastic flow to the rate of settlement. Research is needed in this area.

5.13 Stress–strain and strength characteristics

In order to ensure that construction of the preloading embankment does not cause instability, it is necessary to make a proper assessment of the stress–strain characteristics and shear strength of the foundation.

The effect of the proposed drain installation method on the stress–strain and strength response and on the flow and consolidation characteristics should also be investigated. Good sources of information on laboratory and *in-situ* measurement of shear strength are Schmertmann (1975), Yong and Townsend (1980) and Jamiolkowski *et al.* (1985).

5.14 Secondary compression

In contrast to ordinary foundations on cohesive soils, a proper evaluation of secondary compression may be important in vertical-drain projects, because the time to 100% primary compression is relatively short. Although the magnitude of the post-primary settlement may be much smaller than the primary component, it can be significant if the superstructure is particularly sensitive to settlements. Further, on organic soils and peats, the secondary component can be a major portion of the total settlement.

Practical procedures for estimating the magnitude and rate of secondary compression have been established (Holtz and Kovacs, 1981). Mesri and Godlewski (1977) observed that the ratio C_α/C_c is approximately constant for a wide variety of natural soil deposits (Table 20). C_α is the secondary compression index, (or $\Delta_e/\Delta \log t$); C_c is the

Table 20 Values of the ratio of secondary compression index/compression index for natural soils (as modified by Holtz and Kovacs, 1981; from Mesri and Godlewski, 1977)

Soil	*Secondary, compression index/compression index* C_α/C_c
Organic silts	0.035–0.06
Amorphous and fibrous peat	0.035–0.085
Canadian muskeg	0.09 –0.1
Leda clay (Canada)	0.03 –0.06
Post-glacial Swedish clay	0.05 –0.07
Soft Blue Clay (Victoria, BC)	0.026
Organic clays and silts	0.04 –0.06
Sensitive clay, Portland, Maine	0.025–0.055
San Francisco Bay mud	0.04 –0.06
New Liskeard (Canada) varved clay	0.03 –0.06
Mexico City clay	0.03 –0.035
Hudson river (USA) silt	0.03 –0.06
New Haven (USA) organic clay silt	0.040–0.075

compression index. Figure 34 indicates that the C_α/C_c ratios from some Italian soft clays are slightly lower than those observed by Mesri and Godlewski (1977).

Much uncertainty still exists with respect to the evaluation of the settlement after the preloading surcharge is removed. One approach proposed by Ladd (1971) is to use the value of the rate of secondary compression, $C_{\alpha\varepsilon}$ corresponding to the degree of overconsolidation produced in the soil by the surcharge. This procedure is shown in Figure 35.

Present knowledge does not allow firm conclusions about the contribution of drained creep to the settlement of structures on soft-clay deposits.

5.15 Effect of drain installation: disturbance and remoulding

The installation of vertical drains imposes drastic stress changes (and therefore strains) in the surrounding soil. The soil immediately adjacent to the drain is subject to severe disturbance and remoulding (smear) which can influence the consolidation rate. This influence may be taken into account in the design computations (see Chapter 3, Section 3.3) by assuming an annulus or smear zone of completely remoulded soil around the drain. This concept is a simplifying assumption, because from limited experience with the changes in soil properties from pile driving (e.g. Orrje and Broms, 1967) and sand drains (e.g. Holtz and Holm, 1972; Massarsch and Kamon, 1983) the degree of soil remoulding decreases monotonically from the drain surface. For design purposes, it is necessary to make an assumption concerning the extent of the smeared zone, d_s. However, the selection of this dimension is still

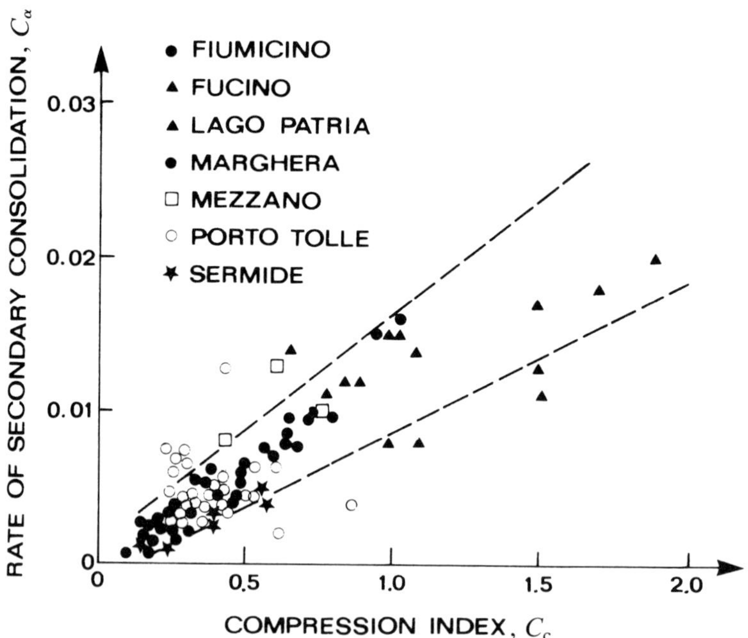

Figure 34 The relationship between compression index and secondary compression for some Italian soft clays (Associazione Geotecnica Italiana, 1979)

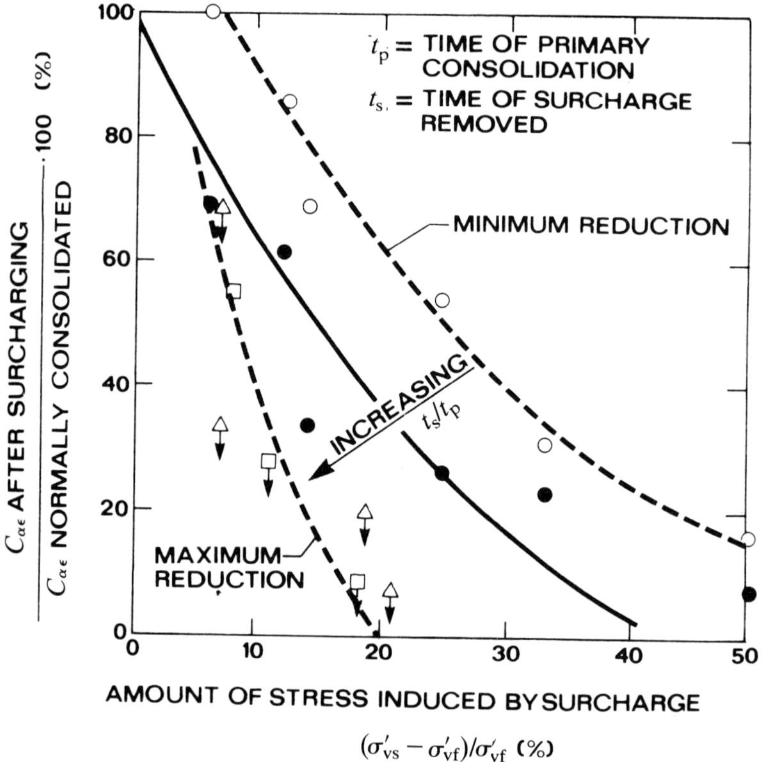

Figure 35 The reduction in rate of secondary compression proposed by Ladd (1971)

somewhat uncertain because of the lack of theoretical and experimental evidence.

Levadoux and Baligh (1980) carried out a comprehensive analysis of the pore pressures developed during cone penetration. They showed that, in undrained conditions, the values of the ratio of initial excess pore pressure to the *in-situ* effective overburden stress $\Delta u_o/\sigma'_{vo}$ are greater than 1.0, even at radial distances of $2d_m$ from the mandrel, where d_m = the diameter of the mandrel used for drain installation. Also, very high octahedral shear strains occurred close to the mandrel, which only decreased to less than 2% beyond a distance of about $2d_m$ from the mandrel.

As noted in Chapter 2, systematic studies of the effects of different installation methods, mandrel shapes, or types of end shoe on smear, disturbance, or field performance have not been made. Even with no macrofabric or anisotropy, disturbance always tends to reduce the value of c_h (see Chapter 3 and Figure 21, page 43) and, as expected, the reduction is more severe with a strongly developed macrofabric and anisotropy. This point deserves additional research.

In the interim, a working hypothesis is to assume that $d_s \simeq 2.5$–$3d_m$. This suggestion is based on the qualitative assumption that the soil is disturbed severely, and thus any anisotropy with respect to its permeability coefficient is almost entirely cancelled, at least to the radial distance at which $\Delta u_o = \sigma'_{vo}$. The value of the remoulded permeability coefficient, k_r, to be assumed within the remoulded zone should be determined experimentally from appropriate laboratory tests. When k_r is not known, conservative estimates may be made from the values of k_v/k_r given in Chapter 3, (Table 11, page 28). These were obtained from laboratory tests on some typical Italian soft clays.

6 Characteristics of drains

The geotechnical engineer, until recently, seldom had the opportunity to select the best filter and drain core for the specific site conditions. There was no choice but to accept the total system as supplied by the manufacturer. But sufficient research results and experience with filters and drains now exist to make it possible to design the drain system and specify the desired material properties for each component. The number of drains currently available and the competition between the various manufacturers, mean that the engineer should be able to obtain the best drain for the geotechnical and loading conditions at the site.

This chapter summarises recent research and experience on the physical characteristics of band-shaped prefabricated drains that affect their installation and performance. The most important performance properties are transverse and longitudinal permeability, as well as the durability of these properties with time. These characteristics relate to the major components of the drain: the filter sleeve and the core. Mechanical properties of drains are important for installation, and they may also play a role in performance. As noted in Chapter 3, Section 3.2, the geometry or shape of drains, particularly their equivalent radius, is important for design purposes.

6.1 Characteristics of the filter sleeve: transverse permeability

The filter sleeves supplied with the drains have characteristics which the manufacturers believe are suitable for a reasonable variety of soil types likely to be encountered in deep-drainage projects. Unfortunately, the filtration characteristics and physical properties of geotextile filter sleeves are often not specified or described clearly, and thus it is difficult to verify whether the filter material is, in fact, appropriate for a specific site.

Nor is it clear from the literature which type of filter is best for prefabricated drains. One school of thought, primarily supported by Hansbo in Sweden, is that the openings in the filter sleeve should be fine enough to prevent soil particles from passing through them (Hansbo, 1981). These fines could cause 'siltation' and reduce the discharge capacity of the drains (Hansbo, 1983a, 1983b). Filter sleeves are quite thin, even if they have a low permeability, and therefore the overall effect on the drain performance is small. This conclusion is based on a theoretical analysis (Hansbo, 1981) in which the filter is considered to be a 'smear zone' of reduced permeability. Hansbo (1983a) believes present drain filters are much too permeable to prevent clogging and siltation of the drain. The only situation in which a high-permeability filter might be an advantage would be where drains are installed in very deep deposits of clay with intermediate sand seams which serve as horizontal drainage layers. The difficulty with the Hansbo approach is that the less permeable filter may have undesirably high head losses which reduce significantly the flow rate especially if with time the permeability reduces below the minimum values suggested by Hansbo (1981).

The other approach, primarily supported by the Dutch on the basis of laboratory experience (den Hoedt, 1981; Jansen and den Hoedt, 1983; Vreeken, van den Berg and Loxham, 1983) is to use a filter with a relatively large open area. This permits the fines to pass through the geotextile filter. The larger soil particles adjacent to it then form a natural soil filter which acts similarly to a graded granular filter (den Hoedt, 1981). Water velocities in the drain are relatively high, at least initially, which means that any soil fines in the drain will probably be flushed out. Therefore, siltation and reduced drain capacity should not occur. Experiments with clays in suspension by Vreeken, van den Berg and Loxham (1983) show that the permeability of drain filter sleeves decreases by a factor of up to 4 or 5, either from clogging of the filter or because of the formation of a filter cake, but this decrease does not appear to have a serious effect on filter performance. Electron micrographs of fine drain filters reported by Kamon (1983) and Kamon and Ito (1984b) indicated very little filter-cake formation, but also that very few fines apparently passed into the drains. Tests indicated a decrease in permeability, but not enough to affect the overall drain-discharge capacity significantly.

Most of the filter sleeves used for prefabricated drains are geotextiles, so it is useful to review the concepts, research results and available experience on the use of geotextiles as filters. This information can then be used to design and specify the materials for drain filter sleeves.

6.2 Geotextile filter concepts

An ideal filter removes or retains all moving soil particles, and at the same time has an infinite permeability, i.e. there should be no head losses or restrictions to flow through it. In geotechnical engineering, an ideal filter usually is not desirable because, in contrast to industrial or sanitary filters, cleaning the filter or removal of an external filter cake is not feasible. In general, the requirements are as follows.

(1) Soil particles should not go into suspension. Therefore, the filter has to prevent piping of the adjacent soil.
(2) Soil particles already in suspension should be allowed to pass through the filter without clogging it.
(3) The filter should have a high permeability so that head losses in the filter itself are negligible.

(4) The filter characteristics should be maintained throughout the life of the project.

These requirements apply both to conventional granular as well as geotextile filters. Additional requirements unique to geotextiles are their tensile strength and resistance to tearing and abrasion to be sufficient to survive installation. The polymer and the type of the geotextile, and their durability properties such as chemical, biological and ultraviolet resistance also have to be considered. Table 21 lists the pertinent properties and test methods for geotextiles in filtration and drainage applications.

Not all the properties in Table 21 have standard tests, although there is much current activity by various American Society for Testing and Materials, British Standards Institution, International Organisation for Standardisation and Réunion International Laboratoire d'Essais Materiaux committees to develop national and international standard test procedures (Gamski, 1984; Cazzuffi, 1986). As noted by

Table 21 Important criteria and properties for filtration and drainage applications (after Bell and Hicks, 1980; Christopher and Holtz, 1985)

Criterion	*Property*	*Source of information or test method*
General	Type and construction	Manufacturer
	Polymer	Manufacturer
	Weight	Manufacturer or ASTM D 1010
	Thickness	Manufacturer or ASTM D 1777*
	Specific gravity; bulk density	Manufacturer or ASTM D 792*
	Absorption (wet weight)	ASTM D 1117
	Cost	Manufacturer
Constructability (survivability)	*Tensile strength*	
	(1) Grab	ASTM D 4633
	(2) Strip	ASTM D 1682, sections 18 and 20 at 300 mm/min
	(3) Wide width	ASTM D 4595
	Tear strength	ASTM D 4533
	Flexibility (stiffness)	ASTM D 1388, Option A*
	Abrasion resistance	ASTM D 4886
	Ultraviolet stability	ASTM D 4355
Hydraulic	*Opening characteristics*	
	(1) Apparent opening size (AOS)	ASTM D 4751
	(2) Pore size distribution (PSD)	Use AOS
	(3) Percent open area	USAE Waterways Experiment Station AD-745-085
	(4) Porosity	No standard test
	Soil retention ability	No standard test – empirical relations to opening characteristics
	Permeability	ASTM D 4491
	Clogging resistance	No standard test; gradient ratio test proposed
	In-plane flow capacity	D 4716
Durability	Clogging resistance	No standard test; gradient ratio test proposed
	Wet and dry stability	No standard test
	Biological resistance	No standard test
	Chemical resistance	No standard test

*Modified – see Christopher and Holtz (1985)

Ingold (1984), such universally accepted standards are essential for the future development of geotextiles and related products.

Geotextile filter concepts have been presented by McGown (1978) and Bell and Hicks (1980). For a geotextile to be an effective filter, it should neither clog nor blind (Figure 36). A geotextile filter can clog if soil particles move and become trapped within the fabric structure. Clogging reduces the permeability of the filter. On the other hand, when moving particles are prevented from entering or passing through the geotextile, a filter can blind. Thus, they coat the surface of the filter to form a filter cake. In this case also, the permeability of the filter can be greatly reduced. With both clogging and blinding, it is possible that the geotextile filter permeability can be even less than that of the surrounding natural soil.

A geotextile filter probably functions in the manner illustrated in Figure 37. A small amount of particle movement occurs into or through the filter leaving the coarser particles to bridge and arch, as shown in the illustration. Sometimes the zone of fine particles immediately behind the soil bridge network is called a filter cake, although it may be difficult to identify in an actual soil because its appearance would be so similar to the natural soil. Once the soil filter zone is established, no further particle movement occurs and the soil–geotextile filter system is in equilibrium. Thus, the geotextile retains the soil filter and prevents its collapse into the drain channels. While it may not actually 'filter' the pore water, the geotextile does act as a catalyst for the formation of a soil filter in the *in-situ* soil. The establishment of a stable and effective self-induced soil filter by the geotextile depends on (McGown, 1976):

(1) The physical and mechanical properties of the geotextile, e.g. the pore size and its distribution, porosity, geotextile thickness and compressibility.
(2) The characteristics of the soil to be protected, e.g. grain size and its distribution, porosity, permeability and cohesiveness.
(3) External stresses and strains imposed on the soil–geotextile system, e.g. traffic and structural loads.
(4) The prevailing hydraulic conditions, e.g. laminar or turbulent flow, unidirectional or reversible flow, dynamic or pulsating flow.

The first attempts to develop geotextile filter criteria followed from experience with graded granular filters. But there are obvious problems with this approach, because the grain size distribution of the geotextile is unknown. It is more appropriate to consider its pore size distribution, but there is no simple way to measure it (Falyse *et*

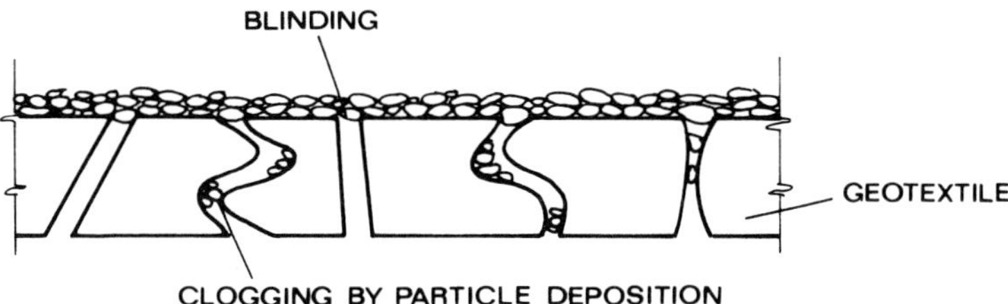

Figure 36 A definition of clogging and blinding (Bell and Hicks, 1980)

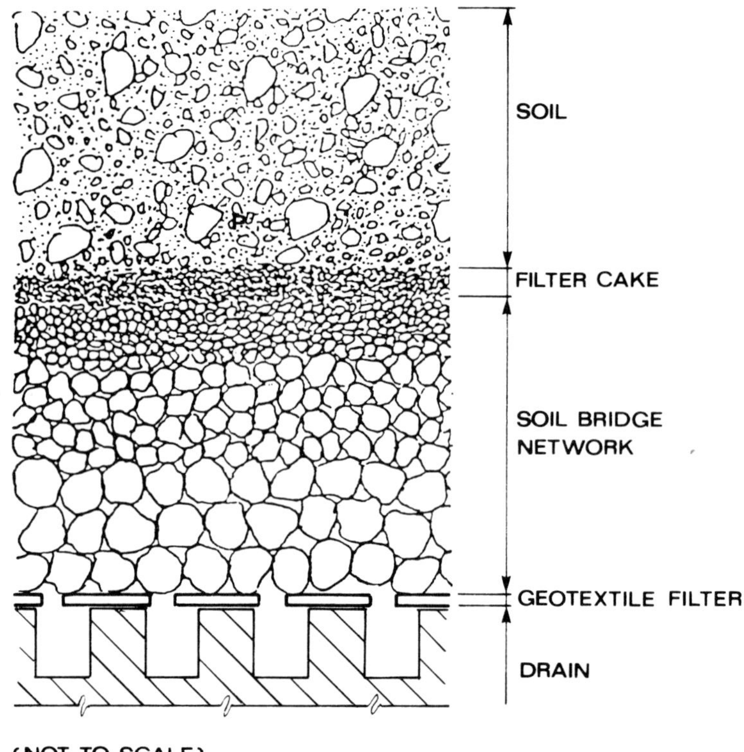

Figure 37 How a geotextile filter probably works (after McGown, 1978 and Bell and Hicks, 1980)

al., 1985; Chen and Chen, 1986) and, as with soils, the relationship between pore size distribution and permeability for geotextiles is not yet established. It may be postulated that a geotextile's permeability, clogging potential and piping resistance depend on its component fibres, its structure (woven, non-woven, composite) and its porosity, but there are no quantitative relationships between these components and permeability or filtration characteristics. This is an important research need.

For woven geotextiles, the size of the openings between the filaments or yarns can be determined with reasonable accuracy. Thus, this opening size, called the apparent opening size, (AOS), has been related to geotextile permeability and filtration characteristics. The numerical AOS value is the opening size of the geotextile in terms of an equivalent British or US Standard sieve for which 95% of the fabric pores are smaller. The symbol for AOS is O_{95}, and corresponds to the sieve opening size in millimetres as determined by the AOS test ASTM D4751 (Amendments) (*American Society for Testing and Materials*, 1989). The AOS is really a measure of the largest hole in the geotextile, and thus is not necessarily a good measure of geotextile permeability or clogging resistance. For non-woven geotextiles (the most common filters for prefabricated drains) the AOS concept is subject to serious criticism (Christopher and Holtz, 1985).

From a detailed study of recent research on the filtration and permeability characteristics of both geotextile and granular filter media, Carroll (1983) draws the following conclusions.

(1) All filter media, granular as well as geotextile, are likely to experience some degree of clogging because of soil infiltration.
(2) The permeability coefficient of a geotextile is not a good indicator of clogging potential.
(3) The AOS of a geotextile is not a good indicator of clogging potential.
(4) Gap-graded soils are prone to soil piping and subsequent filter clogging. High hydraulic gradients also maximise the potential for piping in gap-graded soils.
(5) Well-graded soils are generally not prone to piping. High hydraulic gradients, however, may cause infiltration of even well-graded soils into a filter.
(6) A reasonable limit for maximum allowable gradient ratio is about 3.
The gradient ratio is defined as:

$$\mathrm{GR} = \frac{H_1/L_1}{(H_2 + H_3)/(L_2 + L_3)}$$

where H_1, H_2 and H_3 are the measured heads in the gradient ratio permeameter (Figure 38) and $L_1 = L_2 = L_3 = 25\,\mathrm{mm}$

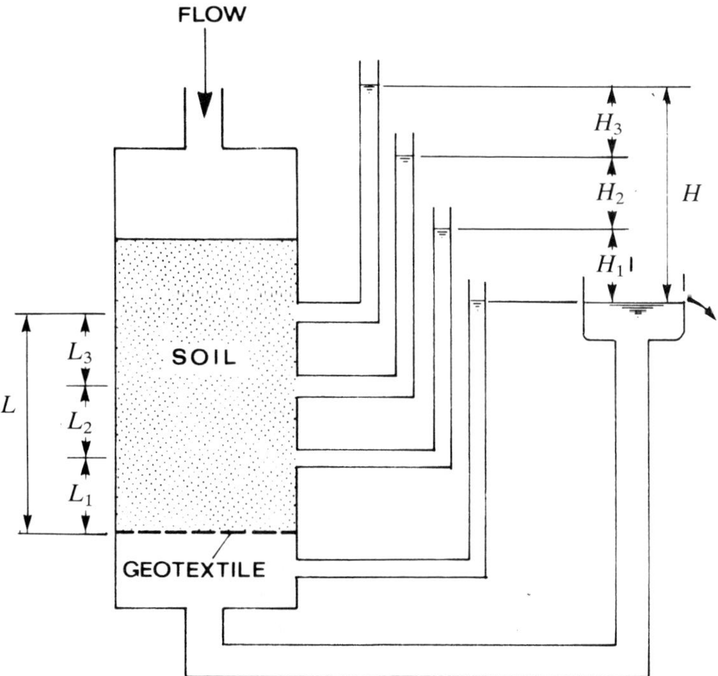

Figure 38 The gradient ratio test

Christopher and Holtz (1985) give details of the gradient ratio test as well as a discussion of some of its problems.

Good summaries of geotextile filtration design techniques developed in both Europe and the USA have been given by Rankilor (1981), Lawson (1982) and Hoare (1982). A design method for geotextile filters based on their recommendations and the research summarised by Carroll (1983) has been developed by Christopher and Holtz (1985) who also suggest testing procedures for determining the geotextile and soil–geotextile properties required for design.

6.3 Designing geotextiles filters* (after Christopher and Holtz, 1985)

A proper design of a geotextile filter requires that the specific soil and hydraulic conditions at a site should be known. The only really effective means to quantify the capability of a particular geotextile filter are by modelling the situation in the laboratory by soil–geotextile performance tests, or by field trials of the proposed soil–geotextile system under the anticipated hydraulic conditions. Quantitative design is particularly important in critical projects and in severe hydraulic conditions. However, it is possible in some situations to evaluate performance qualitatively for less critical applications and less severe hydraulic conditions. A first step in design is to assess how critical to the project is the performance of the drainage and filtration system and how severe are the conditions in which it is to work. Guidelines for this assessment are given in Table 22.

The design procedure presented in the following sections has been adapted to likely vertical-drain situations. It requires a complete engineering evaluation of the project including a knowledge of the soil conditions, flow conditions, and the characteristics of the geotextile filter.

Basically, three performance criteria have to be satisfied: (i) soil retention ability (piping resistance); (ii) permeability, and (iii) clogging resistance.

6.3.1 Soil-retention criteria

The recommended criteria for soil retention (to prevent piping) are as follows.

(1) For steady-state flow:
- (a) woven geotextiles: $O_{95} \leqslant D_{85}$;
- (b) non-woven geotextiles: $O_{95} \leqslant 1.8 D_{85}$;
- (c) geotextile AOS: ($\leqslant 0.3$ mm) ($\geqslant$ no. 50 sieve).

(2) For dynamic, pulsating, or cyclic flow:
- (a) all geotextiles: $O_{50} \leqslant 0.5 D_{85}$

* A committee of the Institution of Civil Engineers has drafted a specificiation for the use of geotextiles and related products and accompanying notes for guidance. These include sections on geotextile filtration. Readers should consult that document, when published, for additional information.

Table 22 Guidelines to assessing the nature of drainage and filtration applications (after Carroll, 1983)

(A) SIGNIFICANCE FOR THE PROJECT: Criterion	*CRITICAL*	*LESS CRITICAL*
(1) Risk of loss of life and/or significant structural damage if drain failure	High	None
(2) Evidence of drain clogging before a potentially catastrophic failure	None	Yes
(3) Repair costs compared with installation cost of drain	Greater	Equal or smaller
(B) SEVERITY OF THE CONDITIONS Criterion	*SEVERE*	*LESS SEVERE*
(1) Soil to be drained	Gap-graded, prone to piping	Well graded or uniform
(2) Hydraulic gradient	High	Low
(3) Flow conditions	Dynamic, cyclic or pulsating	Steady state

In these relationships, O_{95} is the hole size for which 95% of the openings in the geotextile are smaller (it is for practical purposes equal to the AOS, defined in Section 6.2); D_{85} is the particle diameter for which 85% of the particles are smaller; O_{50} is the opening size for which 50% are smaller.

The soil-retention criteria are used to establish an upper bound for the largest allowable opening of the geotextile. Other soil characteristics (such as density), which may contribute to piping, are not included in the criteria. For example, Giroud (1982) has proposed retention criteria for cohesionless soils which include relative density, but such soils are not of interest in deep-drainage applications.

6.3.2 Permeability criteria

The permeability criteria are as follows.

(1) For critical applications and/or severe conditions:

$$k_{\text{geotextile}} \geqslant 10\, k_{\text{soil}}$$

(2) For less critical, less severe situations:

$$k_{\text{geotextile}} \geqslant k_{\text{soil}}$$

In contrast to the retention criteria, the permeability criteria are used to determine the minimum opening-size requirement and minimum fabric porosity. The basic requirement is that the geotextile has to be, and has to remain, more permeable than the adjacent soil, while assuming that some pores clog with time, which effectively reduces the permeability of the fabric. In terms of flow reduction, the criteria are conservative, because the length of flow path through the geotextile filter is very small compared to the soil. Therefore, for less critical and less severe applications, the permeability criterion is less restrictive. In most soil conditions where prefabricated drains are used, geotextiles have more than sufficient permeability, as indicated in Table 23. Test procedures for geotextile permeability are given in Christopher and Holtz (1985) and American Society for Testing and Materials (1989).

An additional criterion for prefabricated drain filters is that their permittivity (flow rate per unit area under a given head) must be as great as the overall flow requirements of the system.

6.3.3 Clogging-resistance criteria

The criteria for clogging resistance depend on the critical nature of the project and on the severity of the hydraulic loading conditions (Table 22).

For critical applications and severe conditions, a geotextile filter should be selected which meets the soil-retention and permeability criteria given above for critical/severe situations. In addition, it should meet the clogging criteria for less-critical/less-severe applications discussed below and soil–geotextile filtration tests using representative samples of the soils at the site should be conducted.

The suggested filtration test for soils with $k > 10^{-7}$ m/s (silts, clayey and silty sands) is the gradient ratio test (see Chapter 6, Figure 38, page 85) and the recommended criterion is $GR \leqslant 3$. For finer grained soils with $k < 10^{-7}$ m/s, long-term filtration tests should be performed.

It should be emphasised that the gradient ratio test has not been fully standardised and that considerable experience is required to obtain reproducible results (Christopher and Holtz, 1985). Filtration tests to measure clogging resistance are soil–geotextile performance tests whose results depend on project and site-specific soil and design conditions. Therefore, it is not reasonable to expect manufacturers to perform them. Such tests should be performed by the specifying agency or its consultant, and they should be conducted prior to selecting or prequalifying the geotextile filter.

An alternative approach for some public works agencies is to use an approved list of geotextiles which have been tested previously and found to be satisfactory. Then, prior to construction, the agency only needs to verify that the filter supplied is, in fact,

Table 23 Typical permeabilities of geotextiles

Geotextile type	*Equivalent Darcy permeability,* k (m/s)*
Monofilament, woven	$10^{-2}-10^{-4}$
Slit-film, woven	$10^{-4}-10^{-5}$
Heat-bonded non-woven	$10^{-2}-10^{-4}$
Resin-bonded non-woven	$10^{-4}-10^{-5}$
Needlepunch non-woven	$10^{-2}-10^{-4}$

*Determined under an effective head of 50 mm ASTM D 4491.

from the approved list. It should be emphasised, however, that laboratory testing is still required for critical projects and to keep the approved list up to date.

For less-critical/less-severe applications, which are probably the majority of vertical-drain projects, the geotextile filter with the largest opening size (smallest AOS number) based on soil retention (Section 6.3.1) should be specified. It is also recommended that, in potential clogging situations such as gap-graded and silty soils, the specification include the following open-area qualifiers.

(1) For non-woven geotextiles: porosity $\geqslant 30\%$.
(2) For woven geotextiles: open area $\geqslant 4\%$.

(This porosity specification is based on graded granular filter porosity requirements.)

Two additional optional qualifiers may also be used:

$$\text{AOS} = O_{95} \geqslant 3D_{15}$$

$$O_{15} \geqslant 2\text{–}3D_{15}$$

where D_{15} = the particle diameter for which 15% of the particles are smaller
O_{15} = opening size for which 15% of the openings are smaller

The first qualifier gives some assurance that: (i) soil fines can, in fact, pass through the largest openings, and (ii) that the minimum pore size is sufficiently large that any moving fines do not clog or blind the filter. These qualifiers are based on experience with grouts in granular soils.

For complete design of the filter, the general, mechanical and durability properties of the geotextile have to be considered (Table 21). Many of these items are discussed in detail later in Sections 6.6, 6.8 and 6.9. The first modern prefabricated drain (Geodrain) had a filter made of paper, and so some additional comments are appropriate. There is no fundamental reason why a paper filter cannot be designed as outlined above, if the appropriate filter characteristics are known. But, it is not as easy to determine these characteristics for papers, mainly because the properties vary with water content and age, and because of the serious problem of durability. Performance characteristics of drains with paper filter sleeves are also discussed in Section 6.8.

6.4 Design examples: geotextile filters

In order to illustrate how the above criteria may be used in practice, examples of the design of filter sleeves for prefabricated drains are given for five different fine-grained soils (A–E). It is assumed that the drainage projects are in the less critical category of Table 22.

The grain-size distribution curves for the five soils are shown in Figure 39 while Table 24 lists the grain-size parameters D_{85} and D_{15} for each soil.

The examples apply the design criteria for soil retention, permeability, and clogging resistance for non-woven filters. The flow is assumed to be steady-state.

Table 24 Grain size parameters for soils in Figure 39

	Soil	D_{85} (mm)	D_{15} (mm)	$1.8D_{85}$ (mm)	$3D_{15}$ (mm)
(A)	Rock flour	0.005	0.0002	0.009	0.0060
(B)	Silty clay	0.010	0.0001*	0.018	0.0003
(C)	Clayey silt	0.050	0.0005*	0.090	0.0015
(D)	Silt	0.075	0.0500	0.135	0.0150
(E)	Organic silt	0.150	0.0150	0.270	0.0450

*Estimated.

Soil retention (piping resistance)

For non-woven filters: $O_{95} \leqslant 1.8D_{85}$ and AOS $\leqslant 0.3$ mm ($\geqslant$ no. 50 sieve). The values for $1.8D_{85}$ are listed in Table 24, and the AOS or O_{95} of the filter specified should be *less* than these values. The second criterion (AOS $\leqslant 0.3$ mm) is met in all cases.

It is probably difficult to find geotextiles to retain soils A and B. This means that if piping occurs, particles move through the filter. Siltation may occur unless the flow rate is sufficient. Siltation and its effect on drain longitudinal permeability is discussed in Section 6.5.3.

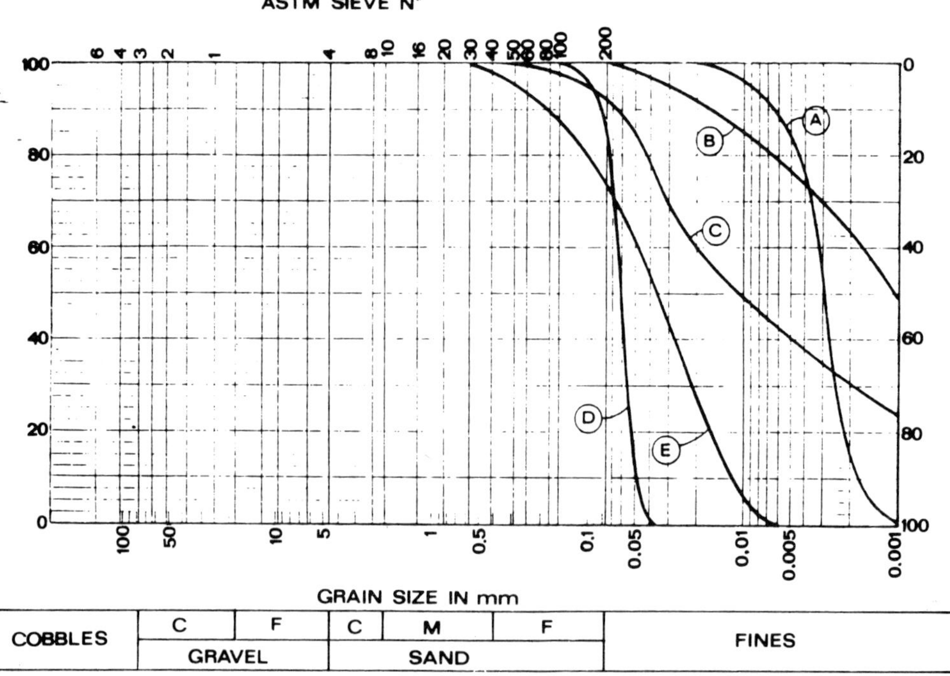

Figure 39 Grain size distribution of soils A–E in the filter design example

6.4.1 Permeability

For less severe conditions, the geotextile permeabilities have to be at least equal to the measured or estimated permeabilities of the five soils. Most geotextiles (Table 23) have sufficient permeability to satisfy this requirement for fine-grained soils.

6.4.2 Clogging potential

For less severe conditions, the retention criteria in Section 6.3.1 should be met. The two additional qualifiers for AOS and O_{15} to the clogging resistance criteria (Section 6.3.3) may also be applied. They are to ensure that the finer soil particles can, in fact, pass through the filter and not clog it. The porosity criterion should also be verified for the silty soils (A, D and E).

Now it is possible to write a specification for the geotextile filter which is best for the soils at each site. But because suitable commercial filters may not be available, some compromises in performance criteria may have to be made.

6.5 Longitudinal permeability: discharge (well) capacity

Once the water has entered the drain, it is still possible for the flow in the drain itself to be reduced for a number of reasons. As noted by Hansbo (1981), resistance to flow in vertical sand drains ('well resistance'), although widely recognised, is usually neglected by designers. The computational complexity required to examine well resistance was thought not to be justified, because of uncertainties in the soil properties and drainage conditions. While these uncertainties still remain, the use of modern electronic computers, together with the simplified equations developed by Hansbo (1981), have effectively eliminated any computational problems.

Hansbo (1981) also showed that, for very long drains, the average degree of consolidation could be reduced significantly by well resistance. The problem with prefabricated drains is the determination of the correct value of the drain discharge capacity, q_w, to be used in design. This value is dependent on a number of factors, such as the following.

(1) The volume available for flow in the core and the effect of the lateral earth pressure on that volume.
(2) Folding and crimping of the drain from large settlements.
(3) Infiltration of fine soil particles through the filter which could cause a reduction of flow capacity of the drain ('siltation').
(4) The durability of the drain system.

These factors are discussed below.

6.5.1 Lateral stress

Jansen and den Hoedt (1983) used the apparatus described by den Hoedt (1981) to test two different types of cores used in a number of prefabricated drains (Figure 40).

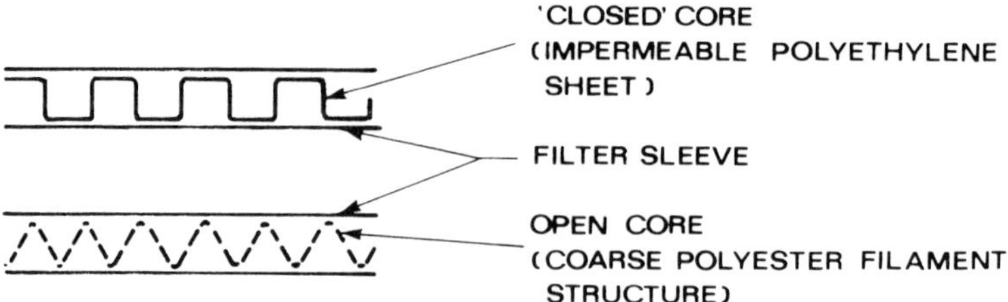

Figure 40 Drain cores tested by Jansen and den Hoedt (1983)

They found that for a uniform silty soil with $k = 10^{-7}$ m/s, the 'open' core had twice as high a discharge capacity, q_w, as that of the 'closed' core but for clayey soils with a lower permeability ($k = 10^{-9}$ m/s), there was little difference in the q_w values. Apparently, with lower-permeability soils, the soil is the dominant factor controlling the flow of water into the drain, not the drain itself. The effect of lateral earth pressure on the discharge capacity depends on several mechanical properties of the filter sleeve and core. In the short term, the extensibility under confining pressure of the filter primarily controls performance. If the filter is relatively extensible (low-modulus) it can be squeezed into the channels and thereby reduce the discharge capacity of the core. Therefore, any measurement of discharge capacity should be performed with the drain embedded in a soil having permeability and stiffness characteristics similar to the soils at the site in question. Further, the measurements should be made under a range of lateral pressures representative of those at the depths of the drains in the field. This effect could develop with time, and therefore the amount and rate are a function of the creep characteristics of the filter sleeve.

Den Hoedt (1981), Kremer *et al.* (1982) and Kremer (1983) performed discharge capacity tests on drains under a range of confining pressures, but their results can be criticised because confinement was provided by either a polyethylene sheet or a rubber membrane and not soil. Suits, Gemme and Masi (1986) also tested various drains under confinement provided by a plastic sheet. Their results suggest that considerable leakage occurred between the plastic sheet and the geotextile filter. On the other hand, limited data presented by Koerner, Fowler and Lawrence (1986) from tests on prefabricated drain cores in the Drexel University lateral-flow device suggest that leakage can be avoided without using soil for confinement. Coincidentally, the Drexel apparatus is very similar to the one in ASTM D4716 in *American Society for Testing and Materials* (1989) for in-plane lateral-flow capacity testing of geotextiles and related materials.

Hansbo (1981, 1983a) published data on q_w of different prefabricated drains under a range of lateral (confining) pressures. These results were in error (Hansbo, 1983b) because of the internal flow resistance of the permeameter device he used. However, the discharge capacity measurements obtained after the reconstruction of this apparatus are in good agreement with those of studies carried out in Milan and reported below.

Kamon and Ito (1984b) confined their drain samples in consolidated kaolin, and their results are shown in Figure 41. They presented similar data from other in-soil

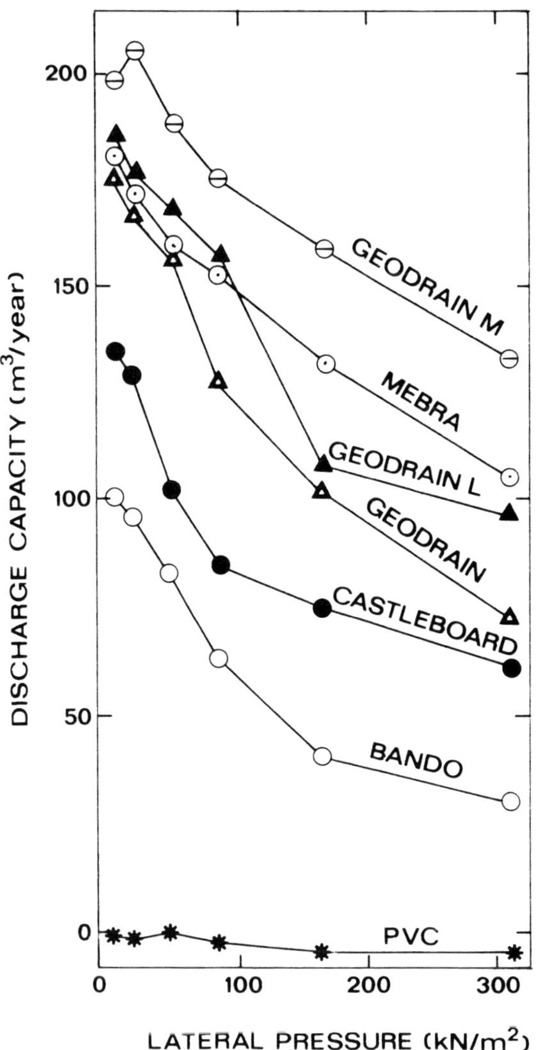

Figure 41 The discharge capacities of prefabricated drains under hydraulic gradient of 0.5 (after Kamon and Ito, 1984a)

tests, in which the void space available for flow decreased for most drains when the confining pressure was increased.

Jamiolkowski, Lancellotta and Wolski (1983b) presented preliminary data on drain discharge capacity tests performed in a large in-soil device (Figure 42) at the laboratories in Milan of Ente Nazionale per l'Energia Elettrica, Centro Ricerca Idraulica e Strutturale (ENEL–CRIS). The drain specimens were placed in soil samples of 300 mm diameter and varying heights of 500–3000 mm. Lateral pressure capacity of the apparatus is about 600 kN/m^2 (equivalent to 40–50 m depth under

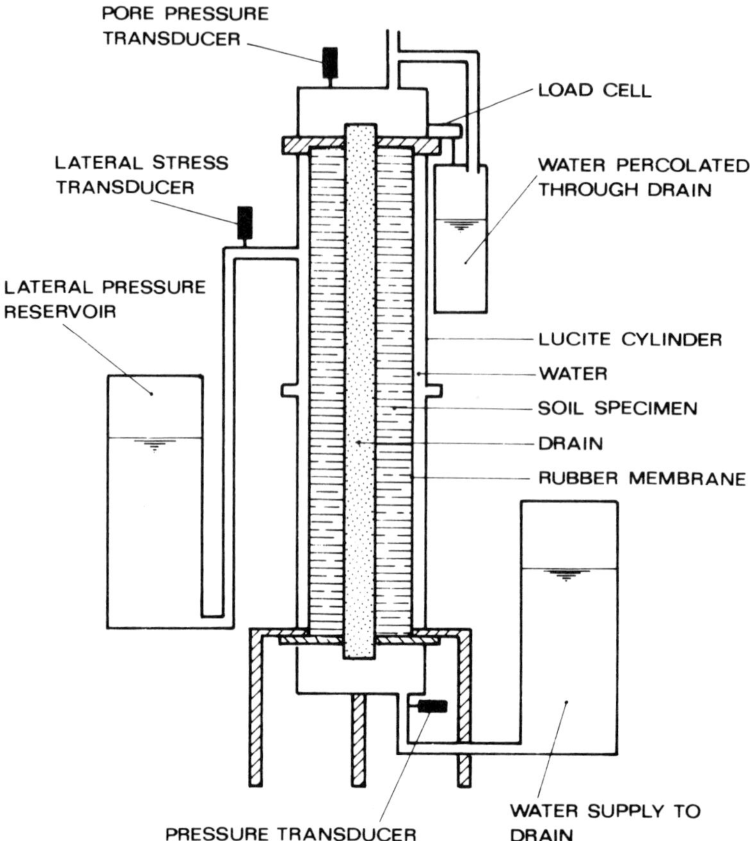

Figure 42 The ENEL–CRIS apparatus for testing prefabricated drains

average conditions). Additional data from ENEL–CRIS (Bellotti and Pedroni, 1986) have been obtained (Figure 43). With only a few exceptions, the discharge capacity of the drains is of the order of several hundred cubic metres per year, which is sufficient for all but the most extreme situations (very long drains, very sandy, silty soils).

The data in Figure 43 have some interesting implications. As can be seen in Figures 43b and 43c, even drains from the same manufacturer can have a wide range of discharge capacities, varying by up to an order of magnitude, depending on the type of core and filter used. Also, there is some variation in results even within the same drain type (Figure 43d). In this respect, the deviation from the mean is ±10–15%. Duplicate tests on other drains had approximately the same variation in results. These results emphasise the importance of basing drain selection on material properties and test values rather than on brand name and model number.

All the tests in Figure 43 were conducted using an hydraulic gradient of 1.0. The effect on discharge capacity of lower hydraulic gradients under a wide range of confining pressures is shown in Table 25. It appears that the drain discharge capacity is not influenced significantly by the hydraulic gradient.

A similar but smaller device and system for testing drains confined either in soil or by a membrane was described by Queyroi, Soyez and Schmitt (1986). Their test results indicated a large decrease in flow capacities when the drains were confined in soil rather than by a membrane only, especially at confining pressures greater than 300 kN/m^2. Overall, their values for discharge capacity, q_w, are similar to those reported above. Chen and Chen (1986) used a similar device to test drains confined in either soil or a rubber membrane. Although presented in a different format from that in Figure 43, the results are similar when their data are converted to the same units. Interestingly, Chen and Chen (1986) found almost exactly the opposite effect of soil confinement to Queyroi, Soyez and Schmitt (1986). The reasons for the differences are not known.

Koda, Szymanski and Wolski (1986) reported the results of some interesting long-term discharge capacity tests on Geodrains. First, five 320 mm diameter and 6 m long perforated steel tubes were installed in peat and organic clay at a test site in the Notec river valley in Poland. Then into each tube Geodrains were inserted in the normal manner. Two Geodrains had non-woven polyester geotextile filters, while the other three had common paper filters. Some 250 and 500 days after installation of the drains, the tubes were extracted. Samples from them, 150 mm diameter and 300 mm long, containing the drains, were trimmed and mounted in a special triaxial-type apparatus for discharge capacity testing. The specimens were reconsolidated to the estimated *in-situ* state of stress, and the discharge capacity of the drains was measured under a hydraulic gradient of 1.0 as the cell pressure was increased incrementally.

The results, shown in Figures 44a for peat and 44b for organic clay, indicate that on a short-term (10 day) basis, a significant decrease in flow rate with increase in confining pressure occurs, but discharge capacities for both paper and geotextile filters were still very satisfactory.

The long-term results were less encouraging, particularly for the paper filter. After 500 days and under a confining pressure of 160 kN/m^2, its flow rate had decreased to 240 m^3/year. A durability problem is suspected in this case, i.e. the paper may be deteriorating because of chemical or biological attack, in addition to its creeping into the drain channels with time. However, the geotextile also showed a decrease in discharge capacity with time, undoubtedly because of creep under stress, but the minimum values observed are somewhat greater than those for the paper filter. Additional comments on durability and long-term resistance of drain materials to chemical and biological attack are presented in Section 6.8.

Den Hoedt (1981) also investigated the behaviour of drains under prolonged pressure but each drain was wrapped in a polyethylene sheet. Thus, the tests probably measured the creep of the filter sleeve material rather than chemical/biological durability. Unfortunately, the test results were not identified with any specific drain or filter type. Kremer (1983) also reported similar tests which indicated a decrease in discharge capacity with time, probably because of filter creep.

Just how much discharge capacity, q_w, is needed? Den Hoedt (1981) stated that the required discharge capacity for a 100 mm wide drain should be at least 10 litres/h

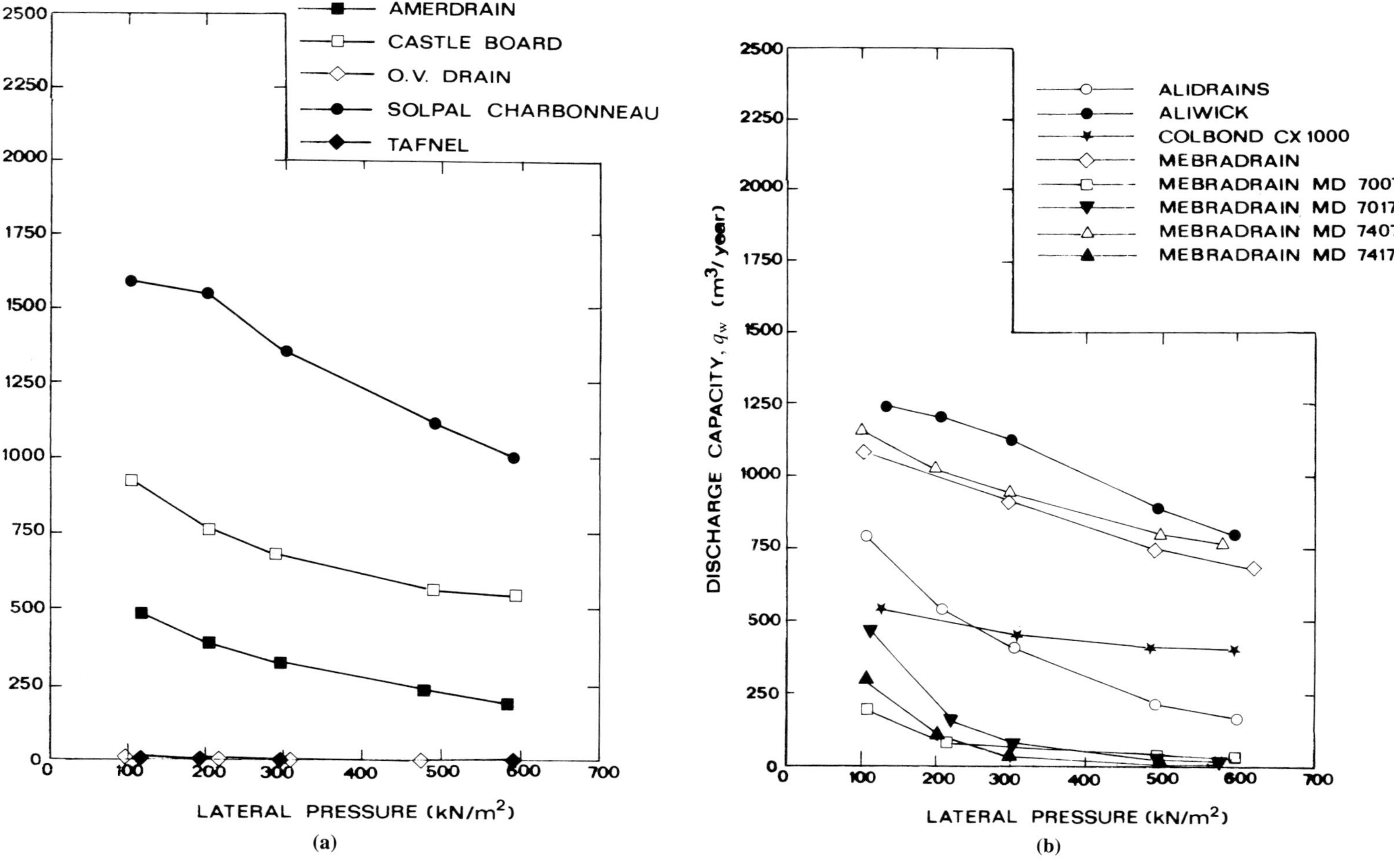

Figure 43 Discharge capacities of several prefabricated drains with a hydraulic gradient of 1.0, performed by ENEL–CRIS

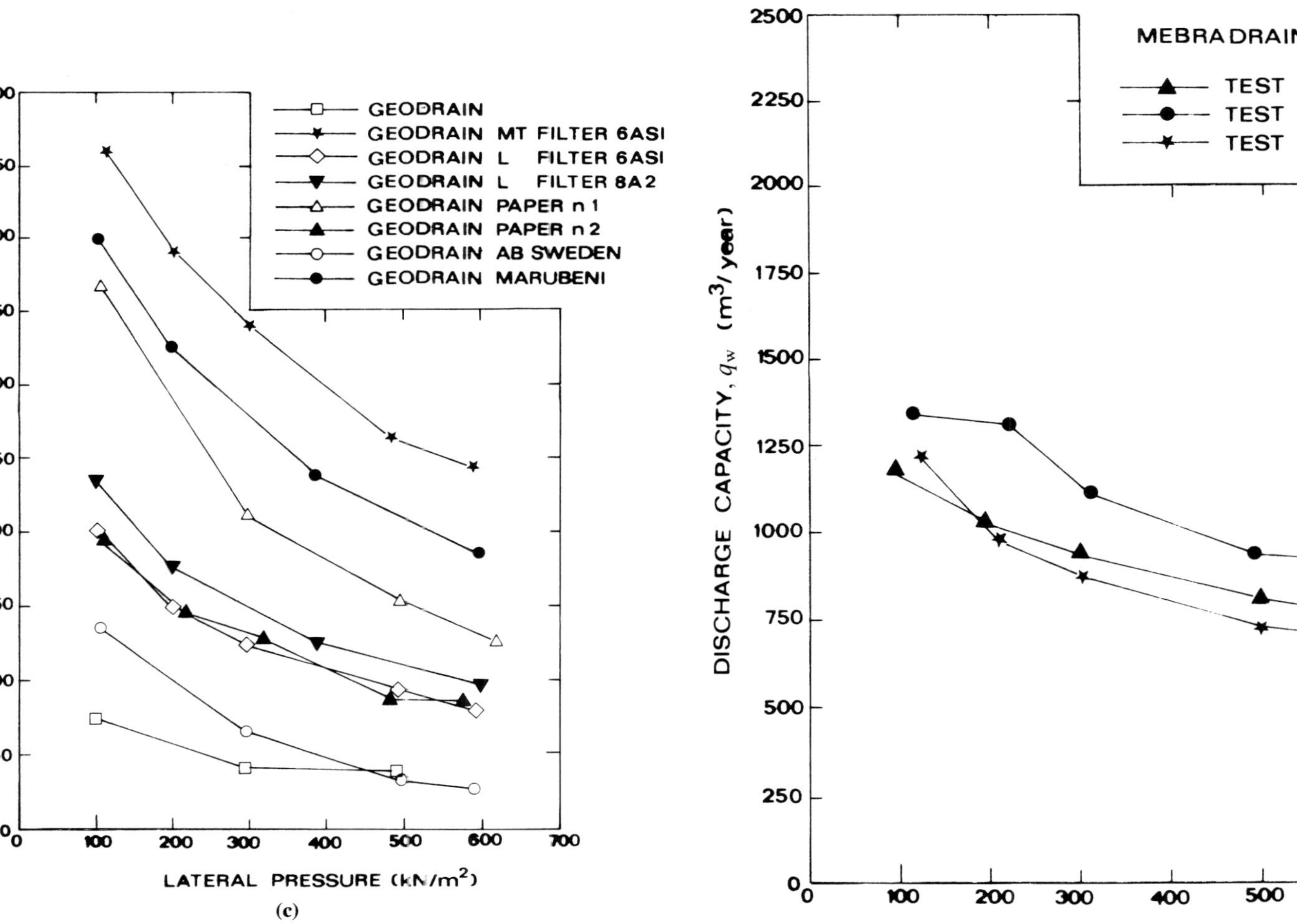

Figure 43 Discharge capacities of several prefabricated drains with a hydraulic gradient of 1.0. performed by ENEL-CRIS (continued)

Table 25 The influence of hydraulic gradient on the measured discharge capacity for four prefabricated drains (data from ENEL–CRIS)

	Lateral pressure *(kN/m²)*	*Hydraulic gradient,* *i*	*Drain discharge capacity,* q_w *(m³/year)*	$\frac{q_w \text{ at } i}{q_w \text{ at } i=1}$
Mebradrain MD 7017	299	1	87	1
	299	0.5	50	0.57
	299	0.2	22	0.25
	492	1	21	1
	492	0.5	13	0.62
	492	0.2	5	0.24
	568	1	18	1
	568	0.5	9	0.50
	568	0.2	3.5	0.19
Mebradrain MD 7417	294	1	41	1
	294	0.5	15	0.37
	294	0.2	4	0.10
	491	1	8	1
	491	0.5	3.6	0.45
	491	0.2	1	0.13
	576	1	5.6	1
	576	0.5	2.9	0.52
	576	0.2	0.9	0.16
Mebradrain MD 7407	99	1	1163	1
	99	0.5	660	0.57
	99	0.2	281	0.24
	295	1	948	1
	295	0.5	506	0.53
	295	0.2	212	0.22
	578	1	768	1
	578	0.5	392	0.51
	578	0.2	154	0.20
Geodrain L (Filter 8A2)	102	1	1095	1
	102	0.5	567	0.52
	102	0.2	229	0.21
	197	1	862	1
	197	0.5	410	0.48
	197	0.2	168	0.20
	294	1	671	1
	294	0.5	341	0.51
	294	0.2	140	0.21

($\sim 3 \times 10^{-6}$ m³/s or 90 m³/year). This requirement is based on an allowable settlement of 40 mm/day for a 30 m long, 100 mm wide, drain. In 1982, the Dutch recommended a minimum q_w of 160 m³/year under a hydraulic gradient, i, of about 0.6 and confining pressure of 100 kN/m² (Kremer *et al.*, 1982). This is the same as their q_w requirement for sand drains. They recommended that testing for discharge capacity should also be performed on a sample of drain folded to a 'flattened-S configuration', and the capacity of the folded drain should be at least 30% of the minimum q_w, or 48 m³/year.

Later, Kremer *et al.* (1982) recommended a minimum drain discharge capacity of

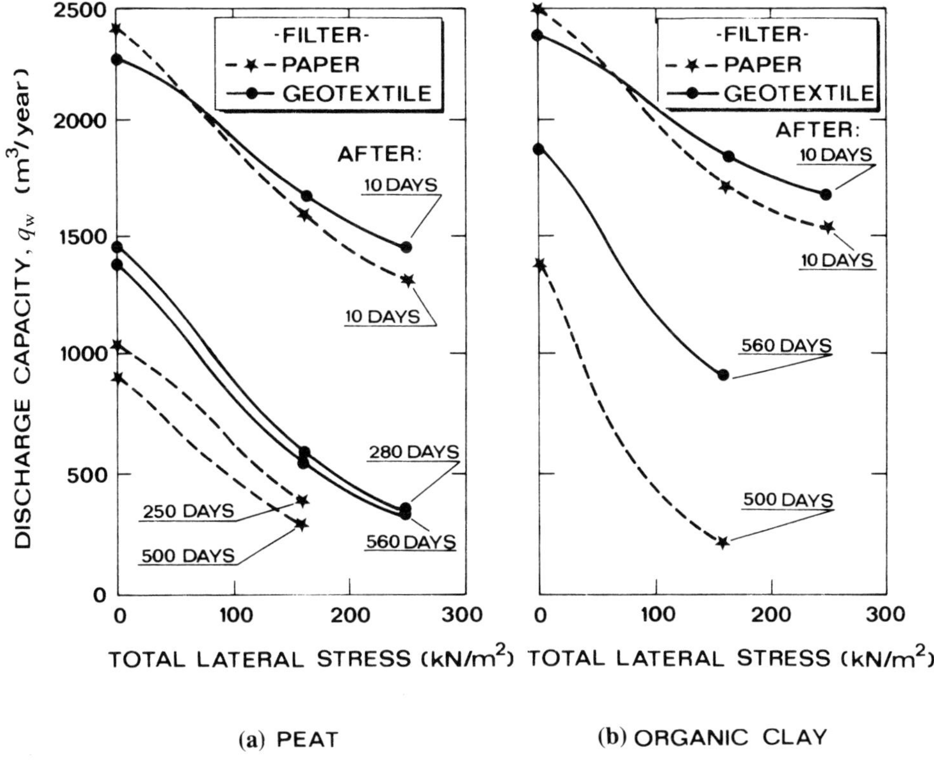

Figure 44 Reductions in the discharge capacities of Geodrains with confining stress and time at a hydraulic gradient of 1.0 (after Koda, Szymanski and Wolski, 1986)

790 m^3/year at 10°C, $i = 1$, and a confining pressure of 15 kN/m^2. This value seems to be unnecessarily high.

On the basis of analytical studies presented in Chapter 3, it is recommended that q_w should have a value of at least 100–150 m^3/year under a confining pressure of 300–500 kN/m^2 to avoid any influence of well resistance. From the data presented in Figure 43 and above, there is no particular problem in meeting this requirement with most currently available prefabricated drains, provided the filters do not deteriorate and the drains do not bend excessively or become clogged with infiltrated soil particles.

6.5.2 Drain folding

Another eventuality which can reduce the discharge capacity of prefabricated drains is bending or folding brought about by large settlements of the clay soils being drained. This possibility is a function of the bending resistance of the drain, which is related to the mechanical properties of the drain and filter. A number of investigators have studied drain folding, both in the laboratory and in the field, but the results are still inconclusive.

Sasaki (1981) reported some preliminary test results which indicated that folding in large-scale laboratory tests did not influence adversely the drain discharge capacity. Kamon and Ito (1984b) showed the shapes of the drains after consolidation of an initially 250mm long section of drain. Unfortunately, neither the amount of settlement which occurred to cause these shapes, nor how much the flow decreased as a result of the 'severe deformation', was reported. Although, in some cases, the drain discharge capacity was greatly reduced, in no case was flow completely shut off. Kremer *et al.* (1982), Kremer (1983) and Oostven (1984) mentioned the possibility of a severe reduction in discharge capacity because of folding from the large vertical strains occurring in very soft soils. They showed a photograph of a severely folded drain in a section of soil, but gave no data on flow rates of folded drains.

Kremer (1983) described a field installation in which drains were excavated and inspected 2 years after installation in a highly compressible peat and clay. The drains were found to be folded to an extent proportional to the settlement of each layer. No water was apparently being discharged from the drains, so an excavation about 5 m deep was made with a back-hoe, and large rings were used to take samples. When the sample in the peaty clay layer was removed, the drain started to flow again. Examination showed a very sharp fold about 50 mm long and only 15mm wide, with the peaty clay pressed completely into the channels of the drain core. In the very sharp fold, the drain core had been totally flattened. Kremer (1983) concluded that folding can reduce the discharge capacity of prefabricated drains to zero, and quality requirements with respect to folding are necessary when relative settlements in the most compressible layers exceed 15%.

Hansbo (1983a) reported that, even after some folding of prefabricated drains in large-scale laboratory tests, no effect of the reduced discharge capacity on the settlement curves was found. Further, Hansbo (1983a) stated that field results do not appear to be affected by folding, probably because its influence is only likely to be significant near the end of the consolidation process when the discharge capcity is of minor importance.

Suits, Gemme and Masi (1985) and Lawrence and Koerner (1988) performed 'crimp' tests, in which short segments of a number of drains were subjected to a 90° bend by a wedge forced into a 90° notch (Figure 45a). Although a number of drains showed a significant reduction in discharge capacity, in no case was flow completely cut off. Futhermore, it was impossible in most cases to predict the bent drain performance based only on the appearance of the core or filter. Tests were conducted at ENEL–CRIS on three prefabricated drains initially bent as shown in Figure 45b during specimen preparation in the apparatus of Figure 42. The results shown in Figure 46 indicate that stiffness and geometry of the filter–core system have a great influence on the resistance of the drain to bending.

If possible bending or kinking of drains is a concern, then flow tests should be conducted using one of the devices shown in Figure 45. Fellenius and Castonguay (1985) observed a slightly different type of folding, which they called 'microfolding', in samples of prefabricated drains recovered from long-term laboratory tests con-

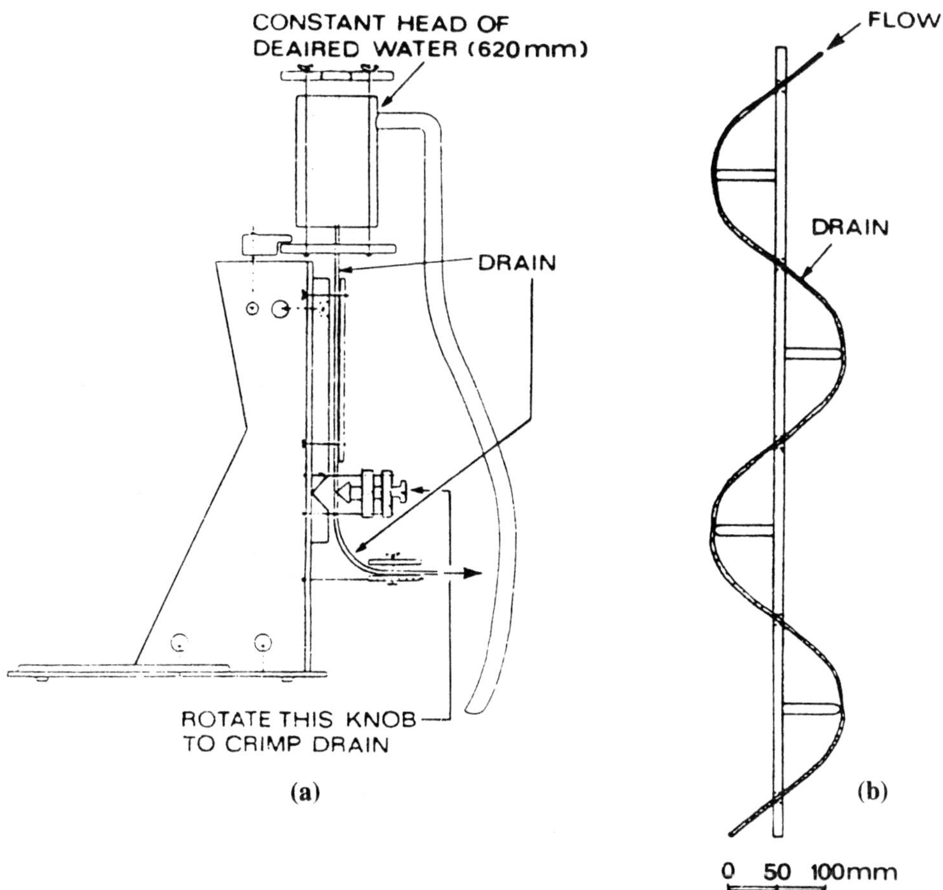

Figure 45 Devices for studying drain folding: (a) the New York State Department of Transportation 'crimp' device (Suits, Gemme and Masi, 1986); (b) ENEL–CRIS test configuration

ducted in 1 m diameter cylinders of soft clay and silt. After consolidation, the vertical compression was 15–20%, and the drains had apparently followed this compression without lateral movement or compression of the drain cores. Instead, the compression was accommodated by: 'contiguous local buckling (microfolding or accordion effect) in the shape of half cylinders with the diameter about equal to the thickness of the drain core'. For cores with studs the folding appeared to have no effect on the volume available for flow. However, for grooved cores, microfolding had deformed the grooves so much that most of the flow volume was lost.

Additional tests were performed at ENEL–CRIS on a number of drains embedded in soft normally consolidated clayey silt in the apparatus shown in Figure 47, after they had been subjected to vertical compression of about 20%. In the straining, the drains became severely folded, resulting in a geometrical configuration of the type shown in Figure 48.

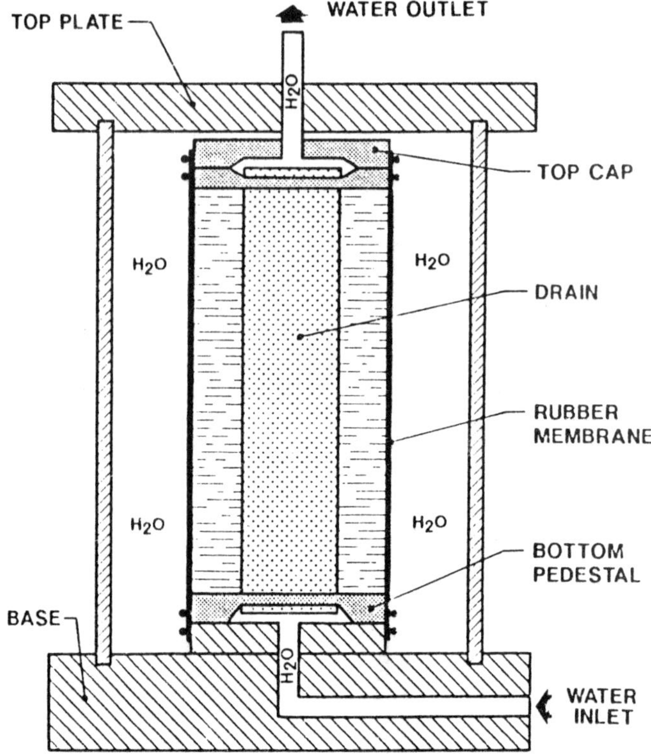

Figure 47 Apparatus by ENEL–CRIS for testing prefabricated drains

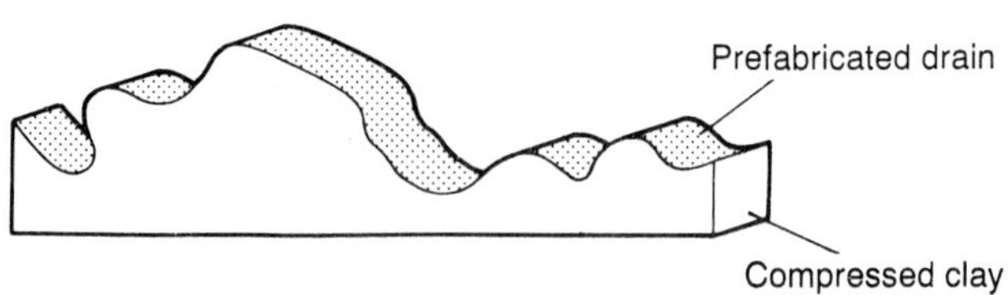

Figure 48 Drains folded after straining

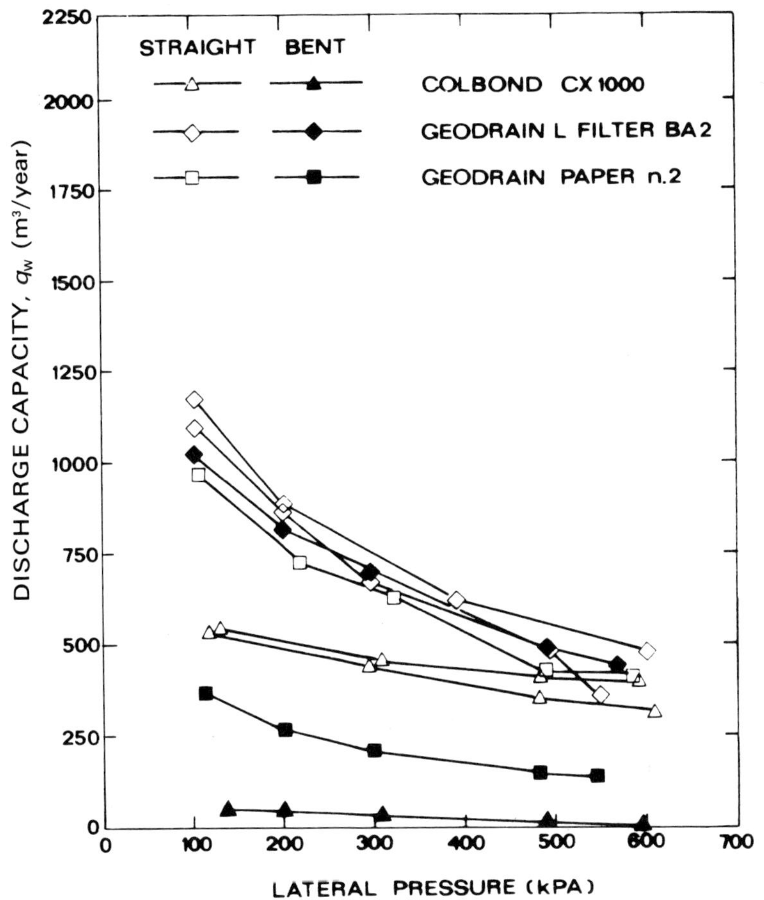

Figure 46 Discharge capacities of straight and initially bent drains in tests by ENEL–CRIS

As a result of the folding the following occurred.

(1) Drains with stiffer plastic cores had reduced discharge capacities, by factors ranging from 5 to 15.
(2) Drains with very flexible plastic cores had only moderate reductions in discharge capacity, by factors ranging between 1.5 and 2.0.

Table 26 presents the results of these tests for a very stiff drain (Colbond, CX 1000) for a stiff drain (Tecnodrain N3) and a flexible drain (Mebradrain). All of them were subjected to 20% vertical compression before the determination of the discharge capacities at different applied horizontal stresses.

In summary, the results of studies of drain folding and its effect on discharge capacity being somewhat inconclusive, additional research on its occurrence and effect is needed urgently.

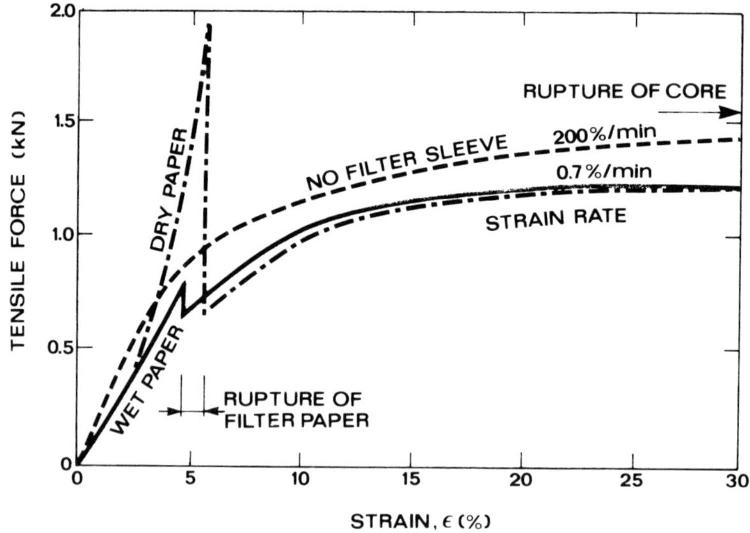

Figure 49 An example of a tension test on a prefabricated drain (after Hansbo, 1983a)

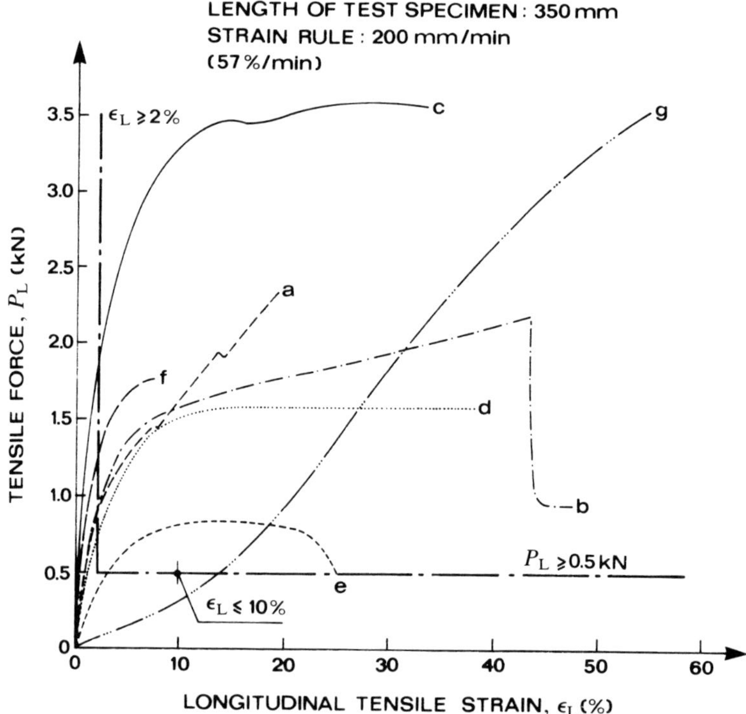

Figure 50 Results of tension tests on complete specimens of seven different (unidentified) drains (after Kremer *et al.*, 1982)

Table 26 Discharge capacities of drains before and after being folded by 25% vertical compression

COLBOND CX 100

Straight		*Folded*	
σ_h (kN/m^2)	q_w (m^3/year)	σ_h (kN/m^2)	q_w (m^3/year)
128	546	104	25
305	456	197	55
482	411	393	50
591	401	586	43

TECNODRAIN 3

Straight		*Folded*	
σ_h (kN/m^2)	q_w (m^3/year)	σ_h (kN/m^2)	q_w (m^3/year)
103	714	106	151
196	560	196	117
296	504	294	125
493	406	492	104
590	379	593	88

MEBRADRAIN

Straight		*Folded*	
σ_h (kN/m^2)	q_w (m^3/year)	σ_h (kN/m^2)	q_w (m^3/year)
99	1462	100	780
196	1214	199	693
295	1042	296	650
492	872	493	507
588	745	589	506

Notes: σ_h=*total lateral stress.*
q_w=*discharge capacity.*

6.5.3 Siltation

Another possibility which could increase well resistance and decrease the discharge capacity is 'siltation', which occurs when very fine soil particles pass through the filter and into the core channels. There may be some circumstances in which the flow rates are sufficient to remove any fines which may have entered the core, e.g. very short drains or fully penetrating drains in which both upward and downward flow occurs. But the more common situation is one in which the hydraulic gradients and flow rates are relatively low and, thus, it is better in these cases simply to prevent infiltration of fines into the core in the first place. As discussed in Sections 6.1–6.4, it should be possible to do this with a properly designed filter sleeve.

6.5.4 Conclusions concerning discharge (well) capacity

Most of the experimental evidence from short-term in-soil tests indicates that well resistance has little or no influence on the consolidation rates of prefabricated drain installations, even for long drains. As long as q_w is greater than 100–150 m^3/year, there should be no significant increase in the consolidation time.

Long-term behaviour is another matter. Filter extensibility and creep under lateral stress mean that the discharge capacity of the core could decrease. Further, deterioration of the filter by clogging, biological degradation or chemical attack is also possible.

Siltation in the drain by fines passing the filter sleeve can be prevented by proper design of the geotextile filter.

Drain bending or folding can possibly decrease the discharge capacity because of large settlements, but the experimental evidence is mixed as to its occurrence in the field and as to its effect. Further research and well-documented field cases are essential in order to find a solution to this problem.

6.6 Mechanical properties of core and filter

The mechanical characteristics of prefabricated drains, particularly the tensile strength of the core and filter, are important because of the stresses to which drains are subjected during installation. These forces are primarily tensile, partly from the drain's self weight and partly from friction between the drain and the installation equipment. According to Kremer *et al.* (1982), the maximum tensile force develops when the mandrel accelerates at the start of the penetration or after slowing because of passing an obstacle or a resistant soil layer.

The drain is pulled from a rotating drum (which usually contains a considerable length of drain) by the mandrel. During penetration, the drain is guided by one or more 'guides'. Thus, the drain has to be able to withstand certain tensile forces in combination with a minimum curvature because of the orientation of the guides. If vibratory equipment is used for installation, the drain is also subject to vibratory forces.

Obviously, the filter should not rupture or tear during installation, and, as noted by Hansbo (1983a), this is best guaranteed by having the stress–strain relationship of the filter coincide, as far as possible, with that of the drain core. The elongation of the filter (at its ultimate stress) and its strength should be as large as possible. Hansbo (1983a) recommends that the stress–strain properties of the drain with a water-saturated filter should be determined in a tension-testing machine. In order to obtain comparable results, the strain rate should be standardised. Perhaps both a low and a high strain rate should be used in order to study the effect of rapid, compared with moderate, installation rates.

Typical examples of tension tests run on prefabricated drain samples under different strain rates are shown in Figures 49 and 50. Chen and Chen (1986) also presented similar results of tension tests on five common drains. All these results suggest that, if installation damage is to be prevented, both the minimum strength and strain required to prevent failure of either component of the drain have to be considered.

Kremer *et al.* (1982) make the recommendations given in the following sections (see also Veldhuijzen van Zanten, 1986).

6.6.1 Tensile strength and modulus of the entire drain

(1) The tensile strength and modulus should be assessed on new (unused) samples of drain with a length of 350 mm.
(2) The rate of strain should be 200 mm/min, i.e. 57%/min.
(3) The longitudinal tensile strength of any of the drain components should be at least 0.5 kN.

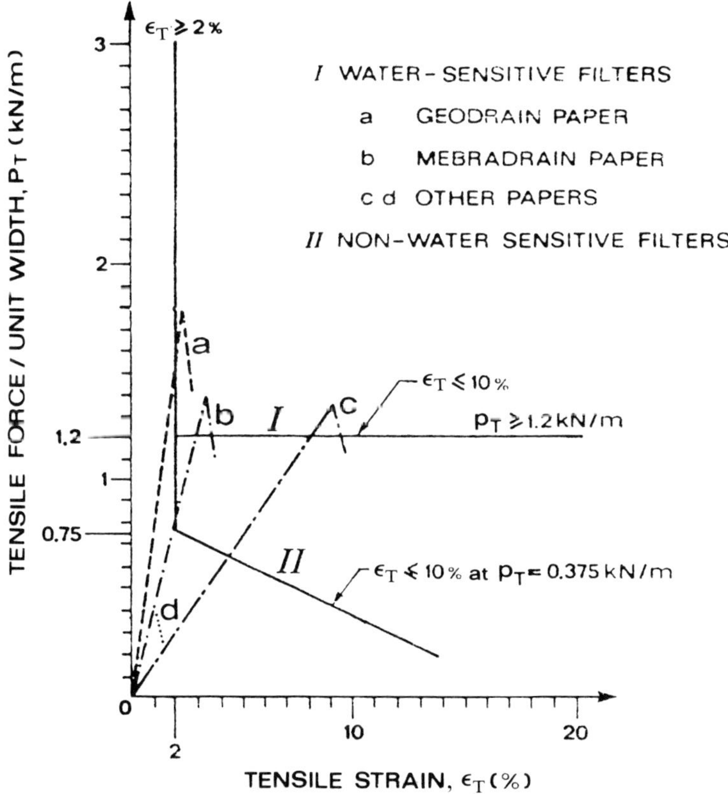

Figure 51 Results of tension tests on wet paper filters (Kremer *et al.*, 1982)

(4) The longitudinal strain at failure should be $\geqslant 2\%$ but $\leqslant 10\%$.
(5) Any seams in the drain filter should have equal or better properties.

The criteria of a tensile load of 0.5 kN and a strain, $\varepsilon_L = 2\%$, are based on estimates of the tensile forces and strains in the drain which may occur during the installation procedure. A tensile test with a strain rate as large as 57%/min is thought to be an accurate simulation of field conditions. The maximum longitudinal strain criterion of $\varepsilon_L = 10\%$ is required in order to limit the deformation of the drain during installation. Large deformations may lead to unwanted decreases in width or thickness of the drain. These criteria seem reasonable in view of the data in Figures 49 and 50.

6.6.2 Tensile strength and modulus of the filter

Different recommendations for the mechanical properties are given depending on whether or not the filters are sensitive to water.

Filters insensitive to water

A sample of filter 175 mm in the width direction of the drain and 50 mm in the length direction should be saturated in water at 10°C for 24 h and tested in widthways tension at a deformation rate of 2 mm/min, i.e. a strain rate of 1.14%/min. The transverse peak tensile strength should be greater than a unit tensile load of 0.750 kN/m at 2% strain or 0.375 kN/m at 10% strain, or any combination linearly interpolated between these values. The transverse strains at failure should be between 2 and 10%.

The criteria for the minimum tensile strength and strain for the filters which are not water-sensitive are based on theoretical considerations and on assumptions of the actual forces and elongations after installation. They are also shown in Figure 51. The limit of transverse strain, $\varepsilon_T = 10\%$ at a transverse load $p_T = 0.375$ kN/m is necessary to prevent reduction of the discharge capacity from the filter sleeve being pressed into the longitudinal channels. This value of p_T is extremely low – almost all geotextiles will have greater tensile strengths (Christopher and Holtz, 1985).

Filters sensitive to water

Tensile tests on filter material should be made prior to wetting, and they should have the following minimum results (Figure 51).

(1) The unit peak tensile strength should be $\geqslant 1.2$ kN/m.
(2) At failure, the strain should be between 2 and 10%.

Filters (such as paper filters) which are sensitive to softening when wet show a decrease in strength and an increase in elongation. However, because they have to remain intact during a certain lifetime, the degree of softening has to be established prior to the drain installation. In the interim, the recommendations of Kremer *et al.* (1982) appear adequate because they were developed for wet paper filters which have shown satisfactory performance in some projects.

Recommended minimum values of physical properties of geotextile filters in drainage and filtration applications to survive construction operations, are given in Table 27. Values listed are interim, and they are based on experience with other geotextile filtration applications, not prefabricated drains. These values are recommended minima to be achieved, which in practice means the actual *minimum average roll values*, not the advertised values given in geotextile manufacturers' brochures.

It is expected that the specific values in Table 27 and the testing requirements will be modified as experience with prefabricated-drain installations increases. It is worth comparing other approaches to specifying geotextile filters (e.g. as in the Institution of Civil Engineers specification for the use of geotextiles and related material, currently in draft for comment), (see footnote, page 86) bearing in mind the several different applications or situations in which geotextile filters are used.

6.7 Geometry and equivalent drain diameter

As discussed in Chapter 3, Section 3.2, the modified theory of consolidation ordinarily used for the design of prefabricated-drain installations assumes radial flow and is written in terms of cylindrical coordinates. The boundary conditions refer to an equivalent outer cylindrical impervious boundary and an inner cylindrical drain. Most prefabricated drains are rectangular in cross-section (band-shaped), and so the rectangular drain has to be converted into an equivalent cylindrical shape yielding the same consolidation rates before the design calculations are made.

Hansbo (1979) suggested that band-shaped, and circular, drains lead to practically the same degree of consolidation, provided their circumferences are equal. Therefore, the equivalent diameter, d_w, of a band-shaped drain with width a and thickness b can be expressed as:

$$d_w = [2(a + b)]/\pi \tag{26}$$

Most prefabricated drains have widths of 95–100 mm and thicknesses of 3–7 mm (see Chapter 2, Table 8, page 11); hence, the equivalent diameter is equal to 62–68 mm.

Jansen and den Hoedt (1983) report that the flow of water to a band-shaped drain is less favourable than that to a cylindrical drain. They suggest, therefore, that a

Table 27 Minimum mechanical properties of geotextile filters for survivability (Christopher and Holtz, 1985)

Property		*Minimum*
Grab strength	(ASTM D 4632) (minimum in either principal direction)	350 N
Puncture strength	(ASTM D 4833)	100 N
Burst strength	(ASTM D 3786)	900 kN/m^2
Trapezoid tear	(ASTM D 4533) (any direction)	100 N

factor of π/4 should be applied, which reduces the equivalent diameter to about 50 mm.

On the other hand, Fellenius and Wager (1977) suggest that the free or open area of the surface of the drain has to be considered because prefabricated drains generally each consist of an open core surrounded by a filter sleeve. Where the sleeve is supported by the core, obviously no water can flow into the drain. They compared the 'open' surface of cylindrical sand drains with prefabricated drains to obtain an equivalent drain diameter of 100–150 mm for the latter. Jansen and den Hoedt (1983) noted that any reduction in the free surface of the drain by the filter sleeve itself is neglected by this theory. Consideration of the porosity of the filter sleeve (which usually varies between 80 and 90%) gives smaller values for the equivalent drain diameter.

Recent laboratory experiments on this point have been inconclusive. From consolidation tests of full-size drains in 1 m diameter cylinders of soft clay and silt, Fellenius and Castonguay (1985) concluded that the equivalent cylinder diameters were 1.5–3.0 times greater than the values determined from Equation (26) for the clay, and 2.5–4.0 times greater for the silt series (85–167 mm for clay and 165–270 mm for the silt tests). For comparison, Equation (26) gives d_w values of 65–68 mm for the drains tested. The range of results occurred because of great uncertainty in the values of c_h and differences in the degree of consolidation based on settlements and that based on pore-pressure observations. There was also a slight difference because the length of time the silt tests were observed was 70 days, compared with 100 days for the clay. Nevertheless, their results verified the earlier conclusion that larger equivalent drain diameters – significantly greater than the value indicated by Equation (26) – are appropriate for commercially available prefabricated drains.

In contrast, tests by Suits, Gemme and Masi (1985) in a 250 mm diameter consolidometer with three different soils showed that the average equivalent diameter for thirteen prefabricated drains ranged between 38 and 64 mm. At the practical drain spacings in the field, this difference is small. For example, the consolidation rate for a 64 mm equivalent sand drain at 1.68 m spacing is the same as that for a 38 mm equivalent sand drain at a spacing of 1.52 m. It is doubtful that drains can be positioned to within 0.16 m.

In conclusion, the recommendation by Hansbo (1979, 1981) simply to use the equivalent circumference, is reasonable at present. Any other refinement seems unnecessary because of the inconclusive research results and other uncertainties in the design parameters.

6.8 Drain durability

When prefabricated band-shaped drains need to be effective for a long time, questions concerning their biochemical and physical resistance arise. Any possible degradation of the filter sleeve which could endanger severely the proper functioning of the drain has to be considered in design. Unfortunately, at present, relatively little is known about this problem.

At the Porto Tolle site in Italy, some 600 000 m of prefabricated drains 28 m long with filter sleeves of non-impregnated paper, were installed during 1975 and 1976 (Garassino *et al.*, 1979; Hegg *et al.*, 1983). Some drains were recovered from the ground 12–15 months after their installation and subjected to careful chemical and physical analyses. The results of these analyses were as follows.

(1) In the uppermost 0.9 m, the filter sleeve had been subjected to microbial degradation, the intensity of which decreased from about 0.2 m down to 0.9 m.
(2) From 0.9 m, down to the maximum explored depth of about 12.3 m, the filter paper was almost intact, but weaker. Locally, at depths of approximately 4 and 23 m, the paper had a black colour from sulphate reduction, probably caused by *Desulphovibria desulphuricus* bacteria. At these locations the filter paper had practically no mechanical strength.

Despite these problems, the overall performance of the prefabricated drains at Porto Tolle ranged from good to acceptable. Settlements ranging from about 1.0 to 2.8 m occurred in time periods between 5 and 10 months (Hegg *et al.*, 1983). Hansbo, Jamiolkowski and Kok (1981) reported on the behaviour of the Porto Tolle trial embankment placed on four different types of drains: Geodrain, Sandwick, Soildrain and 300 mm jetted sand drains (see Chapter 3, Figure 9, page 25). Within a period of 18 months after drain installation, there were no significant differences between the four drain types in the developed settlements and excess pore-pressure dissipation rates. The drain spacings were such that $U = 85\%$ was reached at the same time (Figure 52).

In Basra, Iraq, prefabricated drains with filter sleeves made of impregnated paper have been used. Two years after installation, no traces of paper degradation have been found (Jamiolkowski, Lancellotta and Wolski, 1983a).

The results of a chemical and physical examination of Geodrains with filter sleeves made of non-impregnated paper recovered in 1976 at the Skå-Edeby site in Sweden four years after their installation, showed that – as at the Porto Tolle site – only in the upper 0.6–1.0 m was the paper degraded by microbial activity; below that, it remained mainly intact, even though a strength decrease was apparent. (This information was given to one of the authors by K. E. Eriksson.)

Hansbo (1983b) mentions a very severe degradation of non-impregnated filter paper observed on a drain embedded in the laboratory for 1 year in a sample of organic soil.

Kremer (1983) described an interesting case history in Holland in which prefabricated drains with impregnated paper filters were installed in peaty and clayey soils. Some drains were excavated about 2 years after installation, and the following observations were made.

(1) The impregnated paper filter had disappeared. Only the glue strip remained.
(2) Soil was pressed into the discharge channels of the core and visual inspection indicated that no open space remained in the channels. In the upper layers of the deposit, sand from the fill was found around the drain up to 20 mm thick. Apparently, the extra space created by the mandrel and the anchor plate was filled with sand from the drainage layer.

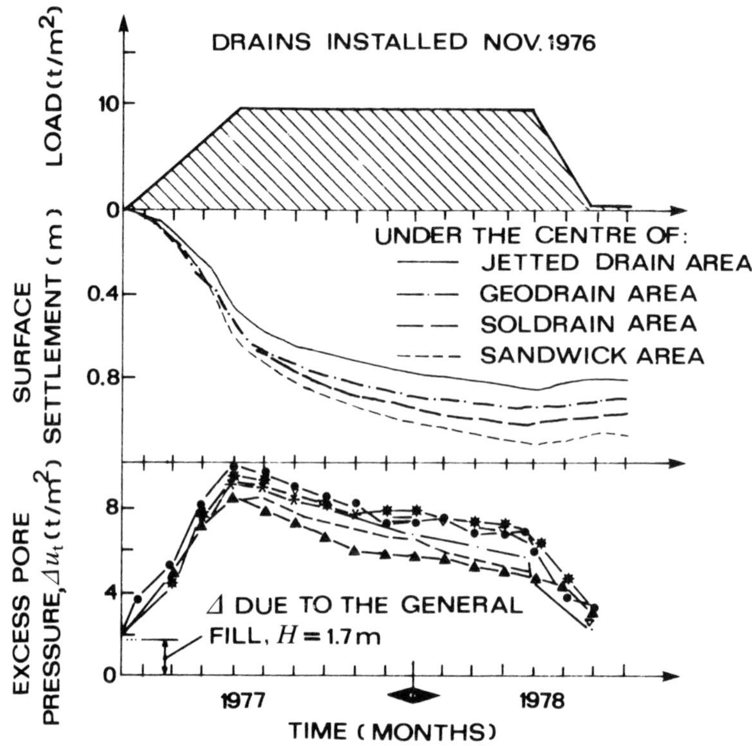

Figure 52 The behaviour of the Porto Tolle embankment on four different types of drain (Hansbo, Jamiolkowski and Kok, 1981)

(3) After drying samples of the drain and surrounding soil in the laboratory, the paper filter was found to be 'reduced to a filmy brittle state with fissures'.

The tests reported earlier by Koda, Szymanski and Wolski (1986) in Section 6.5.1 and Figure 44 showed a marked decrease in the discharge capacity of drains with paper filters after only 250 days. There was also a decrease in the capacity of the drains with a non-woven filter sleeve, but it was not as great as with paper. Part of the decrease in capacity in both types of filters can be attributed to creep of the materials; with the paper filters, both chemical and biological deterioration may have also occurred.

In conclusion, drains with non-impregnated paper filters may be assumed, on the basis of the experience currently available, to be effective for at least 6 (and possibly 12) months, depending on the chemical and bacteriological conditions in the soil. When the filter is made of impregnated paper, the duration may be 24–30 months longer.

There is almost a total lack of well-documented data on the long-term field behaviour of geotextile filter sleeves, particularly in differing soil conditions (Kremer *et al.*, 1982). Veldhuijzen van Zanten (1986) states that 'microbiological activity is of minor importance due to the lack of oxygen' at the common installation depths, but

Table 28 The resistance of various polymers to different agents, by E. W. Cannon (Rankilor, 1981)

Property	*Terylene*	*Nylon 66*	*Nylon 6*	*Nomex*	*Poly-ethylene*
Resistance to:					
fungus	1	3	3	3	4
insects	2	2	2	2	4
vermin	2	2	2	2	4
mineral acids	3	2	2	2	4
alkalis	2	3	3	3	4
dry heat	3	2	2	4	2
moist heat	2	3	3	3	2
oxidising agents	3	2	2	3	1
abrasion	4	4	4	3	3
ultraviolet light	4	3	3	3	1

Property	*Poly-propylene*	*Teflon*	*Polyvinyl chloride*	*Viscous rayons*	*Jute*
Resistance to:					
fungus	3	3	3	3	1
insects	2	3	3	2	1
vermin	2	3	3	3	1
mineral acids	4	4	3	1	1
alkalis	4	4	3	4	1
dry heat	2	4	2	2	2
moist heat	2	4	2	1	2
oxidising agents	3	4	—	1	—
abrasion	3	3	4	3	3
ultraviolet light	3	4	4	2	1

Key to table:
1 Poor
2 Fair
3 Good
4 Excellent

no confirmation is provided. However, there is some information from the performance of geotextiles in other situations. As discussed by Sotton (1984) and Christopher and Holtz (1985), the basic polymers commonly used to make geotextiles are relatively resistant to most detrimental factors found in normal geotechnical environments (Figure 53 and Table 28). In certain situations, such as hazardous-waste ponds and high-pH environments, the geotextile may be exposed to chemical or biological activities which could influence drastically its filtration properties or its durability. The specific site conditions should be reviewed, and if potentially detrimental conditions exist, adjustments to the piping resistance, permeability criteria, clogging resistance and physical property requirements have to be made. Trial exposure to specific detrimental environments may be appropriate, as has been suggested by international standards committees (Cazzuffi, 1986).

Typical durability specifications for geotextile filters should require that the fibres used in their manufacture consist of long-chain synthetic polymers composed of at least 85% by weight of polyolefins, polyesters or polyamides. These polymers should be stabilised to resist deterioration from ultraviolet exposure, and the geotextile filter should not be exposed to ultraviolet radiation (sunlight) for a total of more than 30 days in the period from manufacture to installation. If non-stabilised or highly ultraviolet-susceptible geotextiles are used, they should not be exposed to ultraviolet radiation for more than 5 days.

6.9 Evaluation and specifications of drains

One of the more difficult problems facing the design engineer is the evaluation of new drains as they come onto the market. At present, there are no standard tests available for the measurement of the drain characteristics and properties important for design and selection.

Consequently, the engineer has to rely on manufacturers' claims and/or perhaps on improvised evaluation testing, or only on intuition as a guide in writing drain performance specifications. The proposal by Kremer *et al.* (1982) and Kremer (1983) for international specifications for prefabricated drains is an excellent 'first draft' attempt to bring some order to this problem. Some of the provisions will be adjusted as experience is gained with different drains and installation systems. Rixner, Kraemer and Smith (1986) and Holtz and Christopher (1987) provide a sample specification for prefabricated drains.

McGown and Hughes (1981) discuss a number of practical considerations concerning specifications and construction control of prefabricated drain projects. Hansbo (1983b) has some interesting practical comments on the design, specification and construction with prefabricated drains, as does Choa (1985) in the broader context of soil and site improvement projects. These three contributions merit careful consideration by designers contemplating the use of prefabricated drains.

Procedures for prefabricated drain testing and evaluation which could be the basis for international standards testing include proposals by den Hoedt (1981), Hansbo (1983a), Fellenius and Castonguay (1985) and the approach used in the ENEL–CRIS studies, described in Section 6.5.

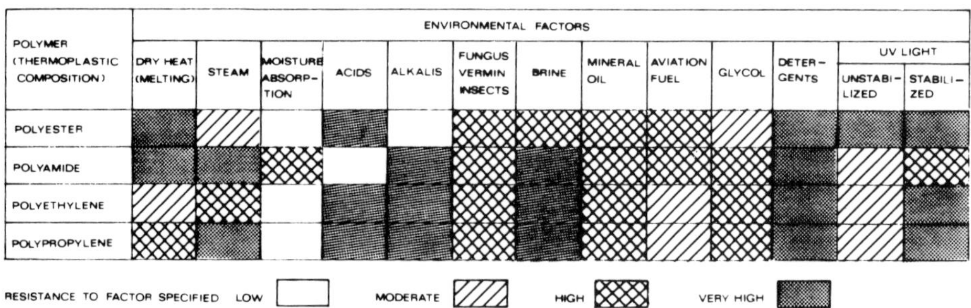

Figure 53 The resistance of common geotextile polymers to environmental factors

7 The determination of design parameters: a status report

The state of the art and practice for the determination of the geotechnical properties at a site and the physical characteristics of prefabricated drains which are required for design have been presented in some detail in this Report. Although it is possible to obtain suitable design values for a number of important properties, there are other areas in which additional research is needed before it is possible to have confidence in all aspects of vertical-drain design. Table 29 is an evaluation of the current state of knowledge. Emphasis is placed on the geotechnical and drain characteristics required for design, rather than on the theoretical/analytical side, because they are major sources of uncertainty in design. With the widespread availability of digital computers, together with the simplified design equations given in Chapter 3 and in the literature, the designer has access to more than adequate analytical capability to design vertical-drain systems (Jamiolkowski, Lancellotta and Wolski, 1983b).

In Table 29, each design topic is rated as: good, satisfactory, or research needed. Improvements are always possible and welcomed for any topic. Nevertheless, in the 'satisfactory' category, our current knowledge is adequate for the vast majority of vertical-drain designs.

For many of the topics, current research shows promise, but it will be some time before its outcome is sufficiently well established for engineers to apply. Table 29, by rating the capability or knowledge available for determining particular design parameters, is more than an identification of research needs. It is also a checklist for the designer of the parameters which have to be estimated and, implicit in the rating, also of those whose evaluation calls for a greater degree of engineering judgement. When the designer is assessing the sensitivity of a vertical-drain design to the choice of parameter values, Table 29 can be used as a starting point. The information given in this Report can then be used by the engineer to judge the reliability of the design.

Table 29 The current capability for designing with prefabricated band-shaped drains (expanded from Koerner, Fowler and Lawrence, 1986)

	Topic	*Current capability*	*Refer to Chapter Section*
(A)	*Theory of vertical drains*		(3)
	(1) Terzaghi–Barron theory	Good	3.7
	(2) Non-linear modifications	Research needed	3.1
	(3) Validity of Darcy's law	Good	3
	(4) Large strain theory	Satisfactory	3.7
	(5) Influence of stress history	Research needed	3.7
	(6) Influence of secondary compression	Research needed	3.7
(B)	*Geotechnical parameters*		(5)
	(1) Consolidation and permeability:		
	(a) laboratory:		5.3
	soils without macrofabric	Satisfactory	
	soils with macrofabric	Research needed	
	(b) field:		5.4, 5.5
	tests in piezometers	Research needed	
	self-boring instruments	Research needed	
	piezocone dissipation tests	Research needed	
	(c) back analysis	Research needed	5.6, 5.7, 5.8
	(2) Stress history:		5.10
	(a) laboratory	Good	
	(b) field: piezocone	Research needed	
	(3) Macrofabric and drainage boundaries:		5.11
	(a) undisturbed sampling	Good	
	(b) *in-situ* testing: piezocone	Satisfactory	
	(4) One- compared with three-dimensional loading	Research needed	5.12
	(5) Stress–strain and strength:		5.13
	(a) laboratory	Satisfactory	
	(b) field	Satisfactory	
	(6) Secondary compression	Research needed	5.14
	(7) Smear, disturbance, installation procedures		2.4, 3.3, 5.15
	(a) extent of disturbed zone	Research needed	
	(b) degree of disturbance and disturbed soil	Research needed	
	properties	Research needed	
	(c) effects of mandrel and end shoe shape	Research needed	
	(d) installation method		
(C)	*Characteristics of drains*		(6)
	(1) Transverse permeability (filter sleeve):		6.1, 6.3
	(a) retention	Satisfactory	
	(b) permeability	Good	
	(c) clogging	Satisfactory	
	(2) Longitudinal permeability (discharge capacity):		3.4, 6.5
	(a) lateral stress	Good	
	(b) drain folding	Research needed	
	(c) siltation	Satisfactory	
	(d) creep of filter into core	Research needed	
	(3) Constructability (survivability):		6.6
	(a) tensile strength of drain	Satisfactory	
	(b) tensile strength of filter	Satisfactory	
	(c) elongation and modulus	Satisfactory	
	(d) tear resistance of filter	Satisfactory	
	(4) Durability:		6.8
	(a) biological	Research needed	
	(b) chemical	Research needed	
	(c) ultraviolet light	Good	
	(d) creep	Research needed	

References

ABOSHI, H. and YOSHIKUNI, H. (1967) 'A study on the consolidation process affected by well resistance in the vertical drain method'. *Soils and Foundations*, **7**, 4, 38–58

ACAR, Y. B., TUMAY, M. T. and CHAN, A. (1982) 'Interpretation of the dissipation of penetration pore pressures', Proceedings, International Symposium Numerical Models in Geomechanics. Zurich

ADACHI, K. and TODO, H. (1979) 'A case study on settlement of soft clay in Penang', Proceedings, Sixth Asian Regional Conference on Soil Mechanics and Foundation Engineering, Volume I, Singapore, pp.117–20

AKAGI, T. (1977a) *Effect of Displacement-type Sand Drains on Strength and Compressibility of Soft Clays.* Dr. Eng. thesis, Department of Civil Engineering, Tokyo University, Tokyo

AKAGI, T. (1977b) 'Effect of mandrel-driven sand drains on strength'. Proceedings, Ninth International Conference on Soil Mechanics and Foundation Engineering, Volume I, Tokyo, pp 3–6.

AKAGI, T. (1979) 'State-of-the-art report on settlements and time rates of consolidation', *Gr. Engng*, **10,** 179–98

ASAOKA, A. (1978) 'Observational procedure of settlement prediction', *Soils and Foundations*, **18,** 87–101 (in English)

AMERICAN SOCIETY FOR TESTING AND MATERIALS (1989) 'Soil and Rock; Building Stones; Geotextiles.' *Annual Book of ASTM Standards*, Volume 04.08, Philadelphia.

ASSOCIAZIONE GEOTECNICA ITALIANA (1979) 'Experienes on the time-settlement behaviour of some Italian soft clays'. Proceedings, Seventh European Conference on Soil Mechanics and Foundation Engineering, Volume 1, Brighton, pp. 1–11.

BALIGH, M. M. (1985) 'Strain path method', *J. Geot. Engr. Div., Am. Soc. Civ. Engrs*, **111,** 9, 1108–36

BALIGH, M. M., AZZOUZ, A. S., WISSA, A. Z. E., MARTIN, R. T. and MORRISON, M. J. (1981) 'The piezocone penetrometer'. Proceedings, American Society Civil Engineers Symposium *Cone Penetration Testing and Experience* (Norris, G. M. and Holtz, R.D. (eds)), St Louis, Missouri, pp.247–63

BALIGH, M. M. and LEVADOUX, J. N. (1980) *Pore Pressure Dissipation after Cone Penetration.* Massachusetts Institute of Technology Research Report R.80–11, Cambridge, Massachusetts

BALIGH, M. M. and VIVATRAT, V. (1979) 'In-situ measurements in a marine clay', Proceedings, *BOSS 79*, Volume I, pp.151–74

BALIGH, M. M., VIVATRAT, V. and LADD, C. C. (1980) 'Cone penetration in soil profiling'. *J. Geot. Engr. Div., Am. Soc. Civ. Engrs*, Vol. 106, **GT4,** 447–61

BARANSKI, T., LECHOWICZ, Z., SZYMANSKI, A. and WOLSKI, W. (1986) 'Field experience with Geodrains in organic soils', Seminar *Laboratory Testing of Prefabricated Band-shaped Drains*, Milan

BARDEN, L. (1965) 'Consolidation of clay with non-linear viscosity'. *Géotechnique*, **15,** 345–62

BARRON, R. A. (1948) 'Consolidation of fine-grained soils by drain wells', *Trans Am. Soc. Civ. Engrs*, **113,** Paper 2346, 718–54

BATTAGLIO, M., JAMIOLKOWSKI, M., LANCELLOTTA, R. and MANISCALCO, M. (1981) 'Piezometer probe test in cohesive deposits', in: Norris, G. M. and Holtz, R. D. (eds), Proceedings, American Society of Civil Engineers Symposium, *Cone Penetration Testing and Experience.* St Louis, Missouri, pp.264–96

BATTAGLIO, M. and MANISCALCO, R. (1983) 'Il Piezocono. Esecuzione ed Interpretazione', *Atti Dell'Istituto di Scienza delle costruzioni del Politecnico di Torino*, 607, Turin.

BEGEMANN, H. K. S. (1966) *The new apparatus for taking a continuous soil sample*, L. G. M. Mededelingen, 4 (Delft), pp.85–104

BELLOTTI, R. and PEDRONI, S. (1986) *Laboratory Testing of Band-shaped Prefabricated Drains.* ENEL-CRIS Internal Report 3394

BELL, J. R. and HICKS, R. G. (1980) *Evaluation of Test Methods and Use Criteria for Geotechnical Fabrics in Highway Applications.* Report to Federal Highway Administration FHWA/RD-80/021. Oregon State University, Corvallis

BENOIT, J. (1983) *Analysis of Self-boring Pressuremeter Tests in Soft Clay.* PhD thesis, Stanford University, California

BIOT, M. A. (1941) 'General theory of three-dimensional consolidation', *J. Appl. Phys.*, **12,** 155–64

BJERRUM, L. (1972) 'Embankments on soft ground', Proceedings American Society of Civil Engineers Conference *Performance of Earth and Earth-supported Structures*, Volume II. West Lafayette, Indiana, pp.1–54

BOWLES, J. E. (1978) *Engineering Properties of Soils and their Measurement.* McGraw-Hill

BROMWELL, L. G. and LAMBE, W. T. (1968) *A comparison of laboratory and field values of* c_v *for Boston Blue Clay. Soils Publication 205 Massachusetts Institute of Technology.* Cambridge, Massachusetts

BRUMMUND, W. F., JONAS, E. and LADD, C. C. (1976) *Estimating* in-situ *Maximum Past Preconsolidation Pressure of Saturated Clays from Results of Laboratory Consolidation Tests in Estimation of Consolidation Settlements*, Transportation Research Board Special Report 163. National Academy of Sciences, Washington DC, pp.4–12

BRUZZI, D. and CESTARI, F. (1983) 'Effect of pore pressure on point resistance using electrical cones in soft clays', Proceedings, International Symposium *Soil and Rock Investigation by* in-situ *Testing*, Paris, Vol. 2, pp.227–30

BURGHIGNOLI, A. (1979) 'Consolidazione monodirezionale e creep delle Argille', *Rivista Italiana di Geotecnica*, **13,** 3, 204–18

CAMPANELLA, R. G. and ROBERTSON, P. K. (1981) *Applied Cone Research.* Soil Mechanics Series 46, University of British Columbia, Vancouver

CARRILLO, N. (1942) 'Simple two- and three-dimensional cases in the theory of consolidation of soils', *J. Math. & Phys.*, **21,** 1–5

CARROLL, R. G. (1983) *Geotextiles Filter Criteria.* Transportation Research Record 916, National Academy of Sciences, Washington, DC, pp 46–53

CARTER, Y. P., RANDOLPH, M. F. and WROTH, C. P. (1979) 'Stress and pore pressure changes in clay during and after the expansion of a cylindrical cavity', *Int. J. Num. & Anal. Meth. Geomech.*, **3,** 305–22

CASAGRANDE, L. and POULOS, S. (1969) 'On the effectiveness of sand drains', *Can. Geotec. J.*, **6,** 3, 287–326

CAZZUFFI, D. (1986) 'Prefabricated drains: testing and material requirements', Proceedings Seminar *Laboratory Testing of Prefabricated Band-shaped Drains.* Milan

CHEN, R. H. and CHEN, C. N. (1986) 'Permeability characteristics of prefabricated vertical drains', Proceedings, Third International Conference, *Geotextiles*, Volume III. Vienna, pp.785–90

CHOA, V. (1985) *Preloading and Vertical Drains.* Proceedings, Third International Geotechnical Seminar, Nanyang Technological Institute, Singapore, pp.87–99

CHOA, V., KARUNARATNE, G. P., RAMASWAMY, S. D., VIJARATNAM, A. and LEE, S. L. (1981) 'Drain performance in Changi marine clay', Proceedings, Tenth International Conference on Soil Mechanics and Foundation Engineering, Volume III, Stockholm, 623–26

CHRISTOPHER, B. R. and HOLTZ, R. D. (1985) *Geotextile Engineering Manual.* U.S. Federal Highway Administration, Washington, DC, *FHWA-TS-86/203*

CLARKE, B. G. (1981) 'In-situ *Testing of Clay using the Cambridge Self-boring Pressuremeter*'. PhD thesis, Cambridge University

CLARKE, B. G., CARTER, J. P. and WROTH, C. P. (1979) 'In-situ determination of the consolidation characteristics of saturated clays', Proceedings, Seventh European Conference on Soil Mechanics and Foundation Engineering, Volume 2, Brighton, pp.207–11

COLLINS, J. T. (1979) 'Laboratory Measurements of Soil Response during Consolidation with Sand Drains.' MSc thesis, University of Texas, Austin, Texas

COUTTS, J. S. (1986) *Correlations between Piezocone Results and Laboratory Data from Six Test Locations.* MSc thesis, University of Surrey

CRAWFORD, C. B. and BURN, K. N. (1976) *Long-term Settlements on Sensitive Clay.* Laurits Bjerrum Memorial Volume, Norwegian Geotechnical Institute, Oslo

D'APPOLONIA, D. J., POULOS, H. G. and LADD, C. C. (1971) 'Initial settlement of structures on clay', *J. Soil Mech. and Found. Div., Am. Soc. Civ. Engrs*, Vol. 97, No. SM10, 1354–77

DARCY, A. (1856) *Les Fontaines Publiques de la Ville de Dijon.* Dalmont, Paris

DAVIS, E. H. (1969) *Rate of Settlement.* Lectures on the analysis of the settlement of foundations. Sydney University, Sydney

DAVIS, E. H. (1971) 'Non-linear consolidation and the effect of layer depth', Proceedings, First Australian and New Zealand Conference *Geomechanics*, Volume I, Melbourne, pp.105–111

DAVIS, E. H. and POULOS, H. G. (1972) 'Rate of settlement under two- and three-dimensional conditions', *Géotechnique*, **22,** 1, 95–114

DEJOSSELINDEJONG, G. (1957) 'Application of stress functions to consolidation problems'. Proceedings, Fourth International Conference on Soil Mechanics and Foundation Engineering, Volume I, London, pp.320–23

DEN HOEDT, G. (1981) 'Laboratory testing of vertical drains'. Proceedings, Tenth International Conference on Soil Mechanics and Foundation Engineering, Volume I, Stockholm, pp.627–30

DE SIMONE, P. and VIGGIANI, C. (1976) 'Consolidation of thick beds of clay', Proceedings, Second International Conference, *Numerical Methods in Geomechanics*, Volume II. Blacksburg, Virginia, pp.1067–81

DROY, D., MCGOWN, A. and MANN, D. C. (1986) 'A8 Port Glasgow Easton section: design and construction', *Proc. Inst. Civ. Engrs*, Part I, **80,** 429–49

FALYSE, E., ROLLIN, A. R., RIGO, J.-M. and GOURC, J.-P. (1985) 'Study of the different techniques used to determine the filtration openings of geotextiles', Proceedings, Second Canadian Symposium on *Geotextiles and Geomembranes* (preprint volume) pp.45–50

FELLENIUS, B. H. and CASTONGUAY, N. G. (1985) *The Efficiency of Band-shaped Drains.* Report to National Research Council of Canada. University of Ottawa, Ottawa, pp.1–54

FELLENIUS, B. H. and WAGER, O. (1977) 'The equivalent sand drain diameter of the band-shaped drain', Proceedings, Ninth International Conference on Soil Mechanics and Foundation Engineering, Volume 3, Tokyo, pp.395–6

FIORAVANTE, V. (1983) *L'Interpretazione delle Prove di Dissipazione Eseguite con il Piezocono.* Tesi Di Laurea, Politecnico di Torino, Turin

FOOTT, R. and LADD, C. C. (1973) *The Behaviour of Altchafalaya Test Embankments during Construction.* Massachusetts Institute of Technology Soil Publication 322, Cambridge, Massachusetts

FURSTENBERG, A., LECHOWICZ, Z., SZYMANSKI, A. and WOLSKI, W. (1983) 'Effectiveness of vertical drains in organic soils', Proceedings, Eighth European Conference on Soil Mechanics and Foundation Engineering, Volume 2, Helsinki, pp.611–616

GAMSKI, K. (1984) 'Classification of geotextiles', Proceedings, Twenty-third Man-made Fibres International Congress *Geotextilien in der Geotechnik*, Dornbirn, Austria, pp.672–88

GARASSINO, A., JAMIOLKOWSKI, M., LANCELLOTTA, R. and TONGHINI, M. (1979) 'Behaviour of pre-loading embankments on different vertical drains with reference to soil consolidation characteristics', Proceedings, Seventh European Conference on Soil Mechanics and Foundation Engineering, Volume 3, Brighton, pp.213–18

GARLANGER, J. E. (1972) 'The consolidation of soils exhibiting creep under constant effective stress', *Géotechnique*, **22,** 1, pp.71–8

GERSEVANOV, N. M. and RATHMAYER, H. (1984) *Recommendations for the rational methods of weak water-saturated subsoil compaction by vertical drains in bases and foundation engineering.* Technical Research Centre Research Note 389, Espoo, Finland. Geotechnical Laboratory and Research Institute of Bases and Underground Structures, Moscow

GHIONNA, V., JAMIOLKOWSKI, M., LANCELLOTTA, R., TORDELLA, M. L. and LADD, C. C. (1981) *Performance of self-boring pressuremeter tests in cohesive deposits.* Massachusetts Institute of Technology Report FHWA/RD-81/173, Cambridge, Massachusetts

GIBSON, R. E., ENGLAND, G. L. and HUSSEY, M. J. L. (1967) 'The theory of one-dimensional consolidation of saturated clays', *Géotechnique*, **17,** 3, 261–73

GIBSON, R. E. and LO, K. Y. (1961) *A Theory of Consolidation for Soils Exhibiting Secondary Compression.* Norwegian Geotechnical Institute Publication 41, pp.1–16

GIBSON, R. E. and MCNAMEE, J. (1957) 'The consolidation settlement of a load uniformly distributed over a rectangular area', Proceedings, Fourth International Conference on Soil Mechanics and Foundation Engineering, Volume I, London, pp.297–99

GIBSON, R. E., SCHIFFMANN, J. K. L. and PU, S. L. (1967) *Plane Strain and Axially Symmetric Consolidation of a Clay Layer of Limited Thickness.* University of Illinois Materials Report, 67-4

GILLESPIE, D. and CAMPANELLA, R. G. (1981) *Consolidation Characteristics from Pore Pressure Dissipation after Cone Penetration.* Soil Mechanics Series N. 47, University of British Columbia, Vancouver

GIROUD, J. P. (1982) 'Filter criteria for geotextiles', Proceedings, Second International Conference on Geotextiles, Volume I, Las Vegas, pp.103–8

HANNA, T. H. (1973) *Foundation Instrumentation.* Transtech Publications, Clausthal

HANNA, T. H. (1985) *Field Instrumentation in Geotechnical Engineering.* Transtech Publications, Clausthal

HANSBO, S. (1960) *Consolidation of Clay, with Special Reference to Vertical Sand Drains.* Swedish Geotechnical Institute, Proceedings No. 18

HANSBO, S. (1979) 'Consolidation of clay by band-shaped prefabricated drains', *Gr. Engng*, **12,** 5, 16–25

HANSBO, S. (1981) 'Consolidation of fine-grained soils by prefabricated drains', Proceedings, Tenth International Conference on Soil Mechanics and Foundation Engineering, Volume 3, Stockholm, pp.677–82

HANSBO, S. (1983a) 'How to evaluate the properties of prefabricated drains', Proceedings, Eighth European Conference on Soil Mechanics and Foundation Engineering, Volume 2, Helsinki, pp. 621–6

HANSBO, S. (1983b) Discussion to Speciality Session 6. Proceedings, Eighth European Conference on Soil Mechanics and Foundation Engineering, Volume 3, Helsinki, pp. 1238–9

HANSBO, S. (1983c) 'Techno-economic trend of subsoil improvement methods in foundation engineering', Special Lectures, Proceedings, Eighth European Conference on Soil Mechanics and Foundation Engineering, Volume 3, Helsinki, pp.1333–43

HANSBO, S., JAMIOLKOWSKI, M. and KOK, L. (1981) 'Consolidation by vertical drains', *Géotechnique*, **31,** 45–66

HARTIKAINEN, J. (1983) 'On the geotechnical design of foundations in improved subsoil', Proceedings, Eighth European Conference on Soil Mechanics and Foundation Engineering, Volume 3, Helsinki, pp. 1319–32

HEAD, K. H. (1982) *Manual of Soil Laboratory Testing*. Pentech Press

HEGG, U., JAMIOLKOWSKI, M., LANCELLOTTA, R. and PARVIS, E. (1983) 'Behaviour of oil tanks on soft cohesive ground improved by vertical drains', Proceedings, Eighth European Conference on Soil Mechanics and Foundation Engineering, Volume 2, Helsinki, pp.627–32

HENKEL, D.J. (1960) 'The shear strength of saturated remoulded clays', Proceedings, American Society of Civil Engineers Conference *Shear Strength of Cohesive Soils*. Boulder, Colorado, pp.533–54

HOARE, D.J. (1982) 'Synthetic fabrics as soil filters: a review', *J. Geot. Engr. Am. Soc. Civ. Engrs,* **108,** GT10, 1230–45

HÖEG, K., ANDERSLAND, O. B. and ROLFSEN, E.N. (1969) 'Undrained behaviour of quick clay under load tests at Asrum', *Géotechnique*, **19,** 1, 101–15

HOLTZ, R.D. and BROMS, B. B. (1972) 'Long-term loading tests at Skå-Edeby, Sweden', Proceedings, American Society of Civil Engineers Conference *Performance of Earth and Earth-supported Structures*, Volume I, Part I. West Lafayette, Indiana, pp.435–64

HOLTZ, R. D. and CHRISTOPHER, B. R. (1987) 'Characteristics of prefabricated drains for accelerating consolidation'. Proceedings, Ninth European Conference on Soil Mechanics and Foundation Engineering, Volume 2, Dublin, pp.903–6.

HOLTZ, R.D. and HOLM B. G. (1972) 'Excavation and sampling around some sand drains at Skå-Edeby, Sweden', Proceedings, Sixth Scandinavian Geotechnical Meeting, Trondheim. Norwegian Geotechnical Institute, pp.75–89. (Also in Reprints and Preliminary Reports, Swedish Geotechnical Institute, N.51.)

HOLTZ, R.D. and KOVACS, W.D. (1981) *An Introduction to Geotechnical Engineering*. Prentice-Hall.

HOLTZ, R.D. and LINDSKOG, G. (1972) 'Soil movements below a test embankment', Proceedings, American Society of Civil Engineers Conference *Performance of Earth and Earth-supported Structures*, Volume I, Part I. West Lafayette, Indiana, pp.273–84.

HUGHES, F.H. and CHALMERS, A. (1972) 'Small-diameter sand drains', *Civ. Engrg & Pub. Wks Rev.*, **67,** 788, 3–6

HVORSLEV, M. J. (1949) *Subsurface Exploration and Sampling of Soils for Civil Engineering Purposes*. Waterways Experiment Station, Vicksburg, Mississippi

I.C.I. FIBRES (1981) *A guide to test procedures used in the evaluation of civil engineering fabrics*. Pontypool, U.K., p.7

INGOLD, T. S. (1984) 'The specification of geotextiles in civil engineering', Proceedings, Twenty-third International Man-made Fibres Congress *Geotextilien in der geotechnik*, Dornbirn, Austria, pp.638–50

JAMIOLKOWSKI, M., LADD, C.C., GERMAINE, J. T. and LANCELLOTTA, R. (1985) 'New developments in field and laboratory testing of soils', Proceedings, Eleventh International Conference on Soil Mechanics and Foundation Engineering, Volume 1 (Theme Lectures), San Francisco, pp.57–153

JAMIOLKOWSKI, M. and LANCELLOTTA, R. (1981) 'Consolidation by vertical drains. Uncertainties involved in the prediction of settlement rates', panel discussion. Session 1, Proceedings, Tenth International Conference in Soil Mechanics and Foundation Engineering. Volume 4, Stockholm, pp.593–5

JAMIOLKOWSKI, M. and LANCELLOTTA, R. (1984) Embankment on vertical drains. Pore pressures during construction. Proceedings, International Conference *Case Histories in Geotechnical Engineering*, Volume I, St Louis, Missouri, pp.275–78

JAMIOLKOWSKI, M., LANCELLOTTA, R. and TORDELLA, M. L. (1980) 'Geotechnical properties of Porto Tolle normally consolidated silty clay', Proceedings, Sixth Danube European Conference on Soil Mechanics and Foundation Engineering, Volume I, Bulgaria

JAMIOLKOWSKI, M., LANCELLOTTA, R. and WOLSKI, W. (1983a) 'Precompression and speeding up consolidation', Proceedings, Eighth European Conference on Soil Mechanics and Foundation Engineering, Volume 3, Helsinki, pp.1201–26

JAMIOLKOWSKI, M., LANCELLOTTA, R. and WOLSKI, W. (1983b) Summary of discussion to Speciality Session 6. Proceedings, Eighth European Conference on Soil Mechanics and Foundation Engineering, Volume 3, Helsinki, pp.1242–5

JANBU, N. (1965) 'Consolidation of clay layers based on non-linear stress strain.' Proceedings, Sixth International Conference on Soil Mechanics and Foundation Engineering, Volume II, Montreal, pp.83–7

JANSEN, H. L. and DEN HOEDT, G. (1983) 'Vertical drains: *in-situ* and laboratory performance and design considerations in fine soils', Proceedings, Eighth European Conference on Soil Mechanics and Foundation Engineering, Volume 2, Helsinki, pp.647–51

JÉZÉQUEL, J.F. and MIEUSSENS, C. (1975) 'In-situ measurement of coefficients of permeability and consolidation of fine soils', Proceedings, American Society of Civil Engineers Conference *In-situ Measurements of Soil Properties*, Volume I. North Carolina State University, Raleigh, pp.208–24

JOHNSON, S. J. (1970a) 'Foundation precompression with vertical sand drains', Proceedings, American Society of Civil Engineers Conference *Placement and Improvement of Soil and Support Structures*, pp.9–42; also in *J. Soil Mech. and Found. Div. Am. Soc. Civ. Engrs*, 96, **SM1,** 145–57

JOHNSON, S. J. (1970b) 'Precompression for improving foundation soils', *J. Soil Mech. and Found. Div. Am. Soc. Civ. Engrs*, 96, **SM1,** 111–70

JOHNSON, S. J. (1974) 'Analysis and design related to embankments: State of the art presentation', Proceedings, A. Soc. Civ. Engrs. Conference *Analysis and Design in Geotechnical Engineering*, Volume II. University of Texas, Austin, pp.1–48

KAMON, M. (1983) Discussion to Speciality Session 6, Proceedings, Eighth European Conference on Soil Mechanics and Foundation Engineering, Volume 3, Helsinki, pp.1232–4

KAMON, M. and ITO, Y. (1984a) 'Function of band-shaped prefabrication plastic board drain', Proceedings, Nineteenth Japanese National Conference on Soil Mechanics and Foundation Engineering, Tokyo

KAMON, M. and ITO, Y. (1984b) 'Variations of discharging capacity due to filter-logging and deformation of plastic board drain materials', Proceedings, Thirty-ninth Annual Conference of the Japanese Society of Civil Engineers, Volume III, Tokyo

KJELLMAN (1937) referenced by Hansbo (1979)

KJELLMAN, W. (1948) 'Accelerating consolidation of fine grained soils by means of cardboard wicks', Proceedings, Second International Conference on Soil Mechanics and Foundation Engineering, Volume II, Rotterdam, pp.302–5

KJELLMAN, W., CADLING, L. and WAGER, O. (1950) 'Soil sampler with metal foils', *Proc. R. Swed. Geotech. Inst.*, 1, Stockholm, 1–76

KODA, E., SZYMANSKI, A. and WOLSKI, W. (1986) 'Laboratory tests on Geodrains: durability in organic soils', Proceedings, seminar *Laboratory Testing of Prefabricated Band-shaped Drains*, Milan

KOERNER, R. M. and BOVE, J. A. (1983) 'In plane hydraulic properties of geotextiles', *Geotechnical Testing Journal,* ASTM, **6,** 4, 190–5

KOERNER, R. M., FOWLER, J. and LAWRENCE, L. A. (1986) 'Soft soil stabilization study for Wilmington Harbour South Dredge Material Disposal Area.' Miscellaneous Paper GL-86-38, USA & Waterways Experiment Station.

KREMER, R. (1983) Discussion to Specialty Session 6. Proceedings, Eighth European Conference on Soil Mechanics and Foundation Engineering, Volume 3, Helsinki, pp.1235–7

KREMER, R., DE JAGER, W., MAAGDENBERG, A., MEYVOGEL, I. and OOSTVEEN, J. (1982) 'Quality standards for vertical drains', Proceedings, Second International Conference on Geotextiles, Volume II. Las Vegas, Nevada, pp.319–24

KREMER, R. H. J., OOSTVEN, J. P., VAN WEELE, A. F., DE JAGER, W. F. J. and MEYVOGEL, I. J. (1983) 'The quality of vertical drainage', Proceedings, Eighth European Conference on Soil Mechanics and Foundation Engineering, Volume 2, Helsinki, pp.721–6

LADD, C. C. (1971) *Settlement Analyses for Cohesive Soils.* Research Report R71–2, Soils Publication 272. Massachusetts Institute of Technology. Cambridge, Massachusetts

LADD, C. C. (1976) 'Use of Precompression and Vertical Sand Drains for Stabilization of Foundation Soils'. Short course on soil and site improvement. University of California, Berkeley, California

LADD, C. C. (1986) 'Stability evaluation for staged construction of embankment on soft ground', Terzaghi Lecture. To be published in *J. Geot. Engr. Am. Soc. Civ. Engrs*

LADD, C. C. and FOOTT, R. (1974) 'New design procedure for stability of soft clays', *J. Geot. Engr. Am. Soc. Civ. Engrs*, **100,** GT7, 763–86

LADD, C. C., FOOTT, R., ISHIHARA, K., SCHLOSSER, F. and POULOS, H. G. (1977) 'Stress-deformation and strength characteristics: State-of-the-art report', Proceedings, Ninth International Conference on Soil Mechanics and Foundation Engineering, Volume 2, Tokyo, pp.421–94

LADD, C. C., RIXNER, J. J. and GIFFORD, D. C. (1972) 'Performance of embankments with sand drains on sensitive clay', Proceedings, American Society of Civil Engineers Conference *Performance of Earth and Earth-supported Structures*, Volume I, Part I, West Lafayette, Indiana, pp.211–42

LAMBE, T. W. (1951) *Soil Testing for Engineers*. John Wiley

LAMBE, T. W. (1973) 'Prediction in soil engineering', 13th Rankine Lecture *Géotechnique*, **23,** 2, 149–202

LANCELLOTTA, R., MANISCALCO, R. and BATTAGLIO, M. (1981) 'Preconsolidazione dei terreni coesivi mediante precarico e dreni', in: *Atti Istituto Scienza delle Costruzioni*, Politecnico di Torino, **505**, Turin

LARSSON, R. and SÄLLFORS, G. (1986) 'Automatic continuous consolidation testing in Sweden', in: Proceedings, Symposium on consolidation of soils: testing and evaluation, *Consolidation Behaviour of Soils*. American Society for Testing and Materials, Special Technical Publication 892. Fort Lauderdale, Florida, pp.299–328

LAWRENCE, C. A. and KOERNER, C. M. (1988) 'Flow behaviour of kinked strip drains'. Proceedings, American Society of Civil Engineers Symposium on Geosynthetics for Soil Improvement. Nashville, Tenn. Holtz, R. D. (ed.) Geotechnical Special Publication No. 18. ASCE

LAWSON, C. R. (1982) 'Filter criteria for geotextiles: relevance and use', *J. Geot. Engr. Am. Soc. Civ. Engrs*, **108,** GT10, 1300–17

LEE, I. K. (1983) *Stabilization of Soft Soils by Surcharging with Special Reference to PVC Drains*. School of Civil Engineering, New South Wales University, Report R-209, Kensington, NSW

LEMINEN, K. and RATHMAYER, H. (1983) 'Experience on horizontal oedometer tests to determine the parameter c_h', Proceedings, Eighth European Conference on Soil Mechanics and Foundation Engineering, Volume 2, Helsinki, pp.647–51

LEROUEIL, S., TAVENAS, F., MIEUSSENS, C. and PEIGNAUD, M. (1978) 'Construction pore pressures in clay foundations under embankments, Part II: Generalized behaviour'. *Can. Geotec. J.*, **15,** 1, 66–82

LEVADOUX, J. N. and BALIGH, M. M. (1980) *Pore Pressure during Cone Penetration in Clays*. Massachusetts Institute of Technology Research Report R80-15. Cambridge, Massachusetts

LONG, R. P. and CAREY, P. J. (1978) *Analysis of Settlement Data from Sand-drained Areas*. Transportation Research Record 678. National Academy of Sciences, Washington, DC, pp.37–40

MAGNAN, J.-P. (1983) *Théorie et Pratique des Drains Verticaux*, Technique et Documentation. Lavoisier, Paris

MAGNAN, J.-P. and DEROY, J. M. (1980) 'Analyse graphique des tassements observés sous les ouvrages', *Bull. de Liaison Lab. P. et Ch.*, **109,** 45–52

MAGNAN, J.-P., PILOT G. and QUEYROI, D. (1983) 'Back analysis of soil consolidation around vertical drains', Proceedings, Eighth European Conference on Soil Mechanics and Foundation Engineering, Volume 2, Helsinki, pp.653–8

MANDEL, J. (1957) 'Consolidation des couches d'argiles', Proceedings, Fourth International Conference on Soil Mechanics and Foundation Engineering, Volume I, London, pp.360–7

MANDEL, J. (1961) 'Tassements produits par la consolidation d'une couche d'argile de grande épaisseur', Proceedings, Fifth International Conference on Soil Mechanics and Foundation Engineering, Volume I, Paris, pp.733–6

MASSARSCH, K. R. and BROMS, B. B. (1983) 'Soil compaction by vibro wing method', Proceedings, Eighth European Conference on Soil Mechanics and Foundation Engineering, Volume 1, Helsinki, pp. 275–8

MASSARSCH, K. R. and KAMON, M. (1983) 'Performance of driven sand drains', Proceedings, Eighth European Conference on Soil Mechanics and Foundation Engineering, Volume 2, Helsinki, pp.659–62

MCGOWN, A. (1976) 'The properties and uses of permeable fabric membranes'. Proceedings, Residential Workshop in Materials and Methods for Low Cost Road, Rail and Reclamation Works, Leura, Australia, pp.663–710

MCGOWN, A. (1978) 'The properties of non-woven fabrics presently identified as being important in public works applications'. *INDEX 78 Conference Paper*, Brussels pp. 1.1–1.31

MCGOWN, A., GABR, A. W. A. and MCNEIL, N. (1979) 'Predicted and measured behaviour of Clyde alluvia beneath Renfrew motorway stage II embankments', Proceedings, Seventh European Conference on Soil Mechanics and Foundation Engineering, Volume 3, Brighton, pp.231–7

MCGOWN, A. and HUGHES, F. H. (1981) 'Practical aspects of the design and installation of deep vertical drains', *Géotechnique*, **31,** 1, 3–17

MCGOWN, A., MARSLAND, A., RADWAN, A. M. and GABR, A. W. A. (1980) 'Recording and interpreting soil microfabric data', *Géotechnique*, **30,** 4, 417–47

MCNAMEE, J. and GIBSON, R. E. (1960) 'Plane strain and axially symmetric problems of the consolidation of a semi-infinite clay stratum', *Q. J. Mech. Appl. Math.*, **13,** Part II, 210–27

MCNAMEE, J. and GIBSON, R. E. (1963) 'A three-dimensional problem of the consolidation of a semi-infinite clay stratum', *Q. J. Mech. Appl. Math.*, **16**

MEIGH, A. C. (1987) *Cone Penetration Testing: methods and interpretation.* Construction Industry Research and Information Association CIRIA Ground Engineering Report. Butterworths

MESRI, G. (1979) Discussion, Proceedings, Seventh European Conference on Soil Mechanics and Foundation Engineering, Volume 4, Brighton, p.141

MESRI, G. (1985) 'Discussion on new developments in field and laboratory testing of soils', Proceedings, Eleventh International Conference on Soil Mechanics and Foundation Engineering, Volume 5, San Francisco

MESRI, G. and CHOI, Y. K. (1979) 'Excess pore water pressure during consolidation', Proceedings, Sixth Asian Conference on Soil Mechanics and Foundation Engineering, Volume I, Singapore, pp.151–4

MESRI, G. and CHOI, Y. K. (1985) 'The uniqueness of the end-of-primary void ratio – effective stress relationships', Proceedings, Eleventh International Conference on Soil Mechanics and Foundation Engineering, Volume 2, San Francisco, pp.587–90

MESRI, G. and GODLEWSKI, P. M. (1977) 'Time and stress-compressibility interrelationship'. *J. Geot. Engr. Div. Am. Soc. Civ. Engrs*, **103,** GT5, 417–30

MESRI, G. and ROKHSAR, A. (1974) 'Theory of consolidation for clays', *J. Geot. Engr. Div. Am. Soc. Civ. Engrs*, **100,** GT8, 889–904

MIEUSSENS, C. and DUCASSE, P. (1977) 'Mesure en place des coefficients de perméabilité et des coefficients de consolidation horizontaux et verticaux', *Can. Geotec. J.*, **14,** 1, 76–90

MIKASA, M. (1985) 'The consolidation of soft clay', in: *Civil Engineering in Japan* (Japanese Society of Civil Engineers), pp.21–6

MILLER, R. J. and LOW, P. F. (1963) 'Threshold gradient for water flow in clay system', *Proc. Soil Sci. Soc. Am.*, **27,** 6, 605

MILLIGAN, V. (1975) 'Field measurement of permeability in soil and rock', Proceedings, American Society of Civil Engineers Conference *In-situ Measurement of Soil Properties*, Volume II. North Carolina State University, Raleigh, pp.3–36

MITCHELL, J. K. (1976) *Fundamentals of Soil Behaviour*. John Wiley

MITCHELL, K. (1981) 'Soil improvement—State-of-the-Art Report'. Proceedings, Tenth International Conference in Soil Mechanics and Foundation Engineering, Volume 4, pp.509–75

MITCHELL, J. K. and GARDNER, W. S. (1975) 'In-situ measurements of volume change characteristics', Proceedings, American Society of Civil Engineers Conference *In-situ Measurement of Soil Properties* Volume II. North Carolina State University, Raleigh, North Carolina, pp.279–345

NICHOLSON, D. P. (1982) Discussion, Institution of Civil Engineers Conference *Vertical Drains*. Thomas Telford

NOIRAY, L. (1982) *Predicted and Measured Performance of a Soft Clay Foundation under Stage Loading*, MSc thesis, Massachusetts Institute of Technology, Cambridge, Massachusetts

OLSEN, H. W. (1965) 'Deviations from Darcy's Law in saturated clays', *Proc. Soil Sci. Soc. of Am.*, **29,** 2, 135–40

OLSON, R. E. (1977) 'Consolidation under time-dependent loading', *J. Geot. Engr. Div. Am. Soc. Civ. Engrs*, **103**, GT1, 55–9

OLSON, R. E. and DANIEL, D. E. (1979) *Measurement of the Hydraulic Conductivity of Fine-grained Soils* American Society for Testing and Materials Special Technical Publication 746, Fort Lauderdale, Florida

OLSON, R. E. and LADD, C. C. (1979) 'One-dimensional consolidation problems', *J. Geot. Engr. Div. Am. Soc. Civ. Engrs*, **105,** GT1, 11–30

ORLEACH, P. (1983) *Techniques to Evaluate the Field Performance of Vertical Drains.* MSc thesis, Massachusetts Institute of Technology, Cambridge, Massachusetts

ORRJE, O. and BROMS, B. (1967) 'Effects of pile driving on soil properties', *J. Soil Mech. and Found. Div. Am. Soc. Civ. Engrs*, **93,** SM5, 59–73

OOSTVEN, J. P. (1984) 'Inspectie van Opgegraven Verticale Drains in Het Veld met Aanvullend Laboratoriumonderzoek', Rapport No. 235. Laboratorium voor Geotechniek, Technical University of Delft

PADFIELD, C. J. and SHARROCK, M. J. (1983) *Settlement of Structures on Clay Soils.* Construction Industry Research and Information Association, CIRIA Special Publication 27, CIRIA

PARRY, R. H. G. and WROTH, C. P. (1977) 'Shear stress–strain properties of soft clays', state-of-the-art lecture, Proceedings, International Symposium on Soft Clay, Bangkok, in: Brand, E. and Brenne, P. W. (eds) (1981) *Soft Clay Engineering*, Chapter 4. Elsevier

PILOT, G. (1977) 'Methods of improving the engineering properties of soft clay', State-of-the-art Report, Proceedings, International Symposium on Soft Clay, Bangkok, in: Brand, E. W. and Brenner, P. W. (eds) (1981) *Soft Clay Engineering*, Chapter 9. Elsevier

PORTER, O. J. (1936) 'Studies of fill construction over mud flats including a description of experimental construction using vertical sand drains to hasten stabilization'. Proceedings, First International Conference on Soil Mechanics and Foundation Engineering, Volume 1, Cambridge, Massachusetts, pp.229–35

QUEYROI, D., SOYEZ, B. and SCHMITT, P. (1986) 'Mesure en laboratoire de la capacité de discharge de drains plats préfabriqués', Proceedings, Third International Conference on Geotextiles, Volume III, *Geotextiles*, pp.803

RANKILOR, P. R. (1981) *Membranes in Ground Engineering*. John Wiley

RATHMAYER, H. (1979) Discussion, Proceedings, Seventh European Conference on Soil Mechanics and Foundation Engineering, Volume 4, Brighton, pp.305–7

RENDULIC, L. (1936) 'Porenziffer und porenwasserdruck in tonen', *Bauingenieur*, **17,** 559

RENDULIC, L. (1937) 'A fundamental principle of soil mechanics and its experimental verification', *Bauingenieur*, **18,** 459

RICHART, F. E. JR. (1959) 'Review of the theories for sand drains', *Trans. Am. Soc. Civ. Engrs*, **124,** paper 2999, 709–39

RIXNER, J. J. KRAEMER, S. R. and SMITH, A. D. Prefabricated Vertical Drains, Volume 1: Engineering Guidelines: Federal Highway Administration Report FHWA/RD-86/168

ROBERTSON, P. K., CAMPANELLA, R. G., BROWN, P. T. and ROBINSON, K. E. (1986) 'Prediction of wick drain and preload performance using piezometer cone data', Proceedings, Thirty-ninth Canadian Geotechnical Conference, Ottawa (preprint); Soil Mechanics Series Publication 100, University of British Columbia, Vancouver

ROWE, P. W. (1968) 'The influence of geological features of clay deposits on the design and performance of sand drains', *Proc. Inst. Civ. Engrs*, paper 7058-S, 1–72

ROWE, P. W. (1972) 'The relevance of soil fabric to site investigation practice', Twelfth Rankine Lecture, *Géotechnique*, **22,** 2, 195–300

ROWE, P. W. and BARDEN, L. (1966) 'A new consolidation cell', *Géotechnique*, **16,** 2, 162–70

ROY, M., TREMBLAY, M., TAVENAS, F. and LA ROCHELLE, P. (1981) 'Behaviour of sensitive clay during pile driving', *Can. Geotec. J.*, **18,** 1, 67–86

ROY, M., TREMBLAY, M., TAVENAS, F. and LA ROCHELLE, P. (1982) 'Development of pore pressure in quasi-static penetration text in sensitive clay', *Can. Geotec. J.*, **19,** 2, 124—38

SANDBAEKEN, G., BERRE, T. and LACASSE, S. (1986) 'Oedometer testing at the Norwegian Geotechnical Institute', Proceedings, Symposium *Consolidation of Soils: testing and evaluation*. American Society for Testing and Materials Special Technical Publication 892, Fort Lauderdale, Florida, pp.329–53

SASAKI, S. (1981) *Report of the Experimental Test for the Prefabricated Drain Geodrain*. Tokyo Construction Co., Tokyo

SCHIFFMAN, R. L. (1958) 'Consolidation of Soil under Time-dependent Loading and Varying Permeability', Proceedings, Highway Research Board, Volume 37. National Academy of Sciences, Washington, DC, pp.584–617

SCHIFFMAN, R. L., CHEN, A. T. F., JORDAN, J. C. (1969) 'An analysis of consolidation theories', *J. Soil Mech. and Found. Div., Am. Soc. Civ. Engrs*, **95,** SM1

SCHLOSSER, F. and JURAN, I. (1979) 'Design parameters for artificially improved soils', Proceedings, Seventh European Conference on Soil Mechanics and Foundation Engineering, Volume 5, Brighton, pp.197–225

SCHMERTMANN, J. H. (1975) 'Measurement of *in-situ* shear strength', Proceedings, American Society of Civil Engineers Conference *In-situ Measurement of Soil Properties*, Volume II, Raleigh, North Carolina, pp.57–138

SENNESET, K., JANBU, N. and SVANØ, G. (1982) 'Strength and deformation parameters from cone penetration tests', Proceedings, *ESOPT II Second European Symposium in Penetration Testing*, Volume 2, Amsterdam, pp.863–70

SINGH, G. and HATTAB, T. N. (1979) 'A laboratory study of efficiency of sand drains in relation to methods of installation and spacing', *Géotechnique*, **29,** 4, 395–422

SKEMPTON, A. W. (1960) 'Effective stress in soils, concrete and rocks', Proceedings, Conference on *Pore Pressure and Suction in Soils*. Butterworths, pp.4–16

SOTTON, M. (1984) 'Durability of geotextiles', Proceedings, Twenty-third International *Man-made Fibres Congress Geotextilien in der Geotechnik*, Dornbirn; Austria, pp.365–88

SUITS, L. D., GEMME, R. L. and MASI, J. J. (1986) 'Effectiveness of prefabricated drains on laboratory consolidation of remolded soils', Proceedings, Symposium on *Consolidation of Soils: testing and evaluation*. American Society for Testing and Materials Special Technical Publication 892, pp.663–83

SUKLJE, L. (1969) *Rheological Aspects of Soil Mechanics*. Wiley Interscience

TAVENAS, F., CHAPEAU, C., LA ROCHELLE, P. and ROY, M. (1974) 'Immediate settlements of three test embankments on Champlain Clay'. *Can. Geotec. J.*, **11**, 1, 109–41

TAVENAS, F. and LEROUEIL, S. (1980) 'The behaviour of embankments on clay foundations', *Can. Geotec. J.*, **17**, 2, 236–60

TAVENAS, F., LEROUEIL, S. and ROY, M. (1982) 'The piezocone test in clays: use and limitations'. Proceedings, *ESOPT II Second European Symposium on Penetration Testing*, Volume 2, Amsterdam, pp.889–94

TAVENAS, F., TREMBLAY, M. and LEROUEIL, S. (1983) 'Mésure *in situ* de la perméabilité des argiles', Proceedings, International Symposium *Soil and Rock Investigations by* In-situ *Testing*, Volume I, Paris, pp.509–13

TAVENAS, F., JEAN, P., LEBLOND, P. and LEROUEIL, S. (1983) 'The permeability of natural soft clays. Part II: permeability characteristics'. *Can. Geot. J.*, **20**, 4, 645–60

TAYLOR, D. W. (1942) *Research on Consolidation of Clays*. Serial 82. Massachusetts Insitute of Technology, Cambridge, Massachusetts

TAYLOR, D. W. (1948) *Fundamentals of Soil Mechanics*. Wiley.

TAYLOR, D. W. and MERCHANT, W. (1940) 'A theory of clay consolidation accounting for secondary compression', *J. Math. Phys.*, **19**, 167–85

TERZAGHI, K. (1923) 'Die Berechnung der Durchlassigkeitsziffer des Tones aus dem Verlauf der Hydrodynamichen Spannungserscheinungen', Akademie der Wissenschaften in Wien, Mathematisch-Naturwissen-Schaftliche Klasse, *Sitzungsberichte*. Abteilung II, **132**, 3/4, 125–38

TERZAGHI, K. (1925) *Erdbaumechanik auf Bodenphysikalischer Grundlage*. F. Deuticke, Vienna

TERZAGHI, K. (1936) 'Stability of slopes of natural soils', Proceedings, First International Conference on Soil Mechanics and Foundation Engineering, Volume I, Cambridge, Massachusetts, pp.161–5

TORSTENSSON, B. A. (1975) 'Pore-pressure sounding instrument', Proceedings, American Society of Civil Engineers Conference, *In-situ Measurement of Soil Properties*, Volume II. North Carolina State University, Raleigh, pp.48–54

TORSTENSSON, B. A. (1982) 'A combined pore-pressure and point resistance probe', Proceedings, *ESOPT II Second European Symposium on Penetration Testing*, Volume II, Amsterdam, pp.903–8

TRAUTWEIN, S. J. (1980) 'Practical Measurements of the Radial Flow Consolidation Properties of Clay'. MSc thesis, University of Texas, Austin, Texas

TRAUTWEIN, S. J., OLSON, R. E. and THOMAS, R. L. (1981) 'Radial-flow consolidation testing', Proceedings, Tenth International Conference on Soil Mechanics and Foundation Engineering, Volume 1, Stockholm, pp.811–4

TREMBLAY, M. (1983) 'Etude des Techniques de mésure *in-situ* de la perméabilité des argiles'. MS thesis, Université Laval, Quebec, p.483

TUMAY, M. T., BOGGESS, R. L. and ACAR, Y. (1981) 'Subsurface investigations with the piezocone penetrometer', in: Norris, G. M. and Holtz, R. D. (eds), *Cone Penetration Testing and Experience*. American Society of Civil Engineers, St Louis, Missouri, pp.325–41

VAN WEELE, A. F. (1983) 'Soil improvement methods in a dense urban environment', Proceedings, Eighth European Conference on Soil Mechanics and Foundation Engineering, Volume 3, Helsinki, pp.1345–50

VELDHUIJZEN VAN ZANTEN, R. (1986) 'The guarantee of the quality of vertical drainage systems', Proceedings, Third International Conference on *Geotextiles*, Volume II, Vienna, pp.651–5

VESIĆ, A. S. (1972) 'Expansion of cavities in infinite soil masses', *J. Soil Mech. and Found. Div. Am. Soc. Civ. Engrs*, **98**, SM3, 265–90

VIGGIANI, C. (1970) Discussion, *J. Soil Mech. and Found. Div. Am. Soc. Civ. Engrs*, **96**, SM1, 331–4

VREEKEN, C., VAN DEN BERG, F. and LOXHAM, M. (1983) 'The effect of clay–drain interface erosion on the performance of band-shaped vertical drains', Proceedings, Eighth European Conference on Soil Mechanics and Foundation Engineering, Volume 2, Helsinki, pp.713–6

WEBER, W. G. (1969) 'Performance of embankments constructed over peat', *J. Soil Mech. and Found. Div. Am. Soc. Civ. Engrs.*, **75**, SM1, 53–76

WELTMAN, A. J. and HEAD, J. M. (1983) *Site Investigation Manual*. Construction Industry Research and Information Association, CIRIA Special Publication 25, CIRIA

WEST, J. M. (1975) 'The role of ground improvement in foundation engineering', *Géotechnique*, **25**, 1, 71–8

WISSA, E. A., CHRISTIAN, J. T., DAVIS, E. H. and HEIBERG, S. (1971) 'Consolidation at constant rate of strain', *J. Soil Mech. and Found. Div. Am. Soc. Civ. Engrs.*, **97**, SM10, 1393–413

WISSA, A. E. Z., MARTIN, R. T. and GARLANGER, J. E. (1975) 'The piezometer probe', Proceedings, American Society of Civil Engineers Specialty Conference *In-situ Measurement of Soil Properties*, Volume I. North Carolina State University, Raleigh, pp.536–45

WROTH, C. P. (1984) 'The interpretation of *in-situ* soil tests', Twenty fourth Rankine Lecture, *Géotechnique*, **34**, 4, 449–89

YONG, R. M. and TOWNSEND, F. C. (eds) (1981) Symposium *Laboratory Shear Strength of Soil.* American Society for Testing and Materials Special Technical Publication 740

YOSHIKUNI, H. and NAKANODO, H. (1974) 'Consolidation of fine-grained soils by drain wells with finite permeability', *Soils and Foundations*, **14,** 2, 35–46

Index